WAVES AND SATELLITES IN THE NEAR-EARTH PLASMA

STUDIES IN SOVIET SCIENCE

PHYSICAL SCIENCES

A Continuation Order Plan is available for this series. A continuation order will bring delivery of each new volume immediately upon publication. Volumes are billed only upon actual shipment. For further information please contact the publisher.

STUDIES IN SOVIET SCIENCE

WAVES AND SATELLITES IN THE NEAR-EARTH PLASMA

Ya. L. Al'pert

*Institute of Terrestrial Magnetism
and Ionospheric Radio Wave Propagation (IZMIRAN)
Academy of Sciences of the USSR
Moscow, USSR*

Translated from Russian by
Julian B. Barbour

CONSULTANTS BUREAU • NEW YORK AND LONDON

Library of Congress Cataloging in Publication Data

Al'pert, IÂkov L'vovich.
  Waves and satellites in the near-Earth plasma.

  (Studies in Soviet science)
  Translation of Volny i iskusstvennye tela v prizemnoi plazme.
  Includes bibliographical references.
  1. Plasma waves. 2. Ionospheric radio wave propagation. 3. Artificial satellites.
I. Title. II. Series.
QC718.5.W3A4613   530.4'4   74-19475
ISBN 978-1-4684-8476-2   ISBN 978-1-4684-8474-8 (eBook)
DOI 10.1007/978-1-4684-8474-8

This translation was prepared from a copy of the manuscript submitted to the Soviet publisher, as well as additions supplied by the author. A Russian edition of this work will be published shortly in the USSR. This translation is published under an agreement with the Copyright Agency of the USSR (VAAP).

© 1974 Consultants Bureau, New York
Softcover reprint of the hardcover 1st edition 1974

A Division of Plenum Publishing Corporation
227 West 17th Street, New York, N.Y. 10011

United Kingdom edition published by Consultants Bureau, London
A Division of Plenum Publishing Company, Ltd.
4a Lower John Street, London W1R 3PD, England

FOREWORD

This book presents a brief review of the main results obtained in two new branches of plasma physics that have developed rapidly in the last decade following the launching of artificial satellites. The aim has been to illuminate results that have a certain completeness and permanent nature and will retain their significance and be used in further investigations. A further aim has been, as far as possible, to acquaint the reader with the most recent achievements in these interesting branches of modern science.

The first chapter of the book contains some data, theoretical results, and formulas that will be used to consider different types of wave phenomena that occur in the ionosphere, magnetosphere, and the solar wind. The second chapter contains experimental and theoretical results obtained from the study of the flow of plasmas around bodies. Here, theory predominates over experiment, which reflects the state of development of these investigations. The results of the second chapter will undoubtedly retain their significance in the future. The writing of the third chapter presented the most difficult problem. The literature is being continuously augmented with the results of investigations of wave processes that occur in the plasma that is nearest to the Earth -- regions of the ionosphere at an altitude of 200-300 km and more -- out to distances from the Earth of millions of kilometers -- in the solar wind. We shall refer to all this region of plasma as the near-Earth plasma. Here, there is a multitude of good experimental studies and new interesting results. Numerous attempts are made to explain them theoretically. However, the majority of theoretical studies in recent years do not give a direct quantitative, and frequently not even persuasive indirect explanation of the experimental facts. This is due in a number of cases to the absence of an adequate set of basic experimental data, the complexity of the problems that must be solved, which frequently require the investigation of nonlinear equations, or rather the finding of ad-

equate nonlinear mechanisms for the processes. As a result, the greater part of the theoretical studies is
methodological and sometimes speculative in nature. The
problems are not solved "head on", as one requires for
a quantitative, unambiguous explanation of the experimental data. At the same time, the majority of the known
experiments can, on closer examination, be discussed in
the framework of phenomena that have been well studied
in linear plasma theory and are briefly described in part
in the first chapter. In view of the brevity of this
review and this state of the theory, it appeared advisable to include in the third chapter only the main
results of the diverse experiments. They characterize,
in the whole range of frequencies, almost all wave modes
observed in a natural plasma. Experimental facts have
been selected that, in my opinion, will retain their significance for future investigations and can be regarded
as the foundations and basic data of this branch of plasma physics.

I hope that this book will be of interest to many
readers and in particular the specialists that work in
these branches of physics.

I should like to take this opportunity of thanking
A. V. Gurevich, who read the manuscript of the book care-
fully, for a number of comments. I am also grateful to
N. I. Bud'ko, A. P. Dubovoi, and A. M. Moskalenko for
individual comments, and L. I. Bud'ko and V. L. Morozova
for assistance in preparing the manuscript.

CONTENTS

Contents

WAVES AND SATELLITES IN THE NEAR-EARTH PLASMA

INTRODUCTION

Two new directions of investigation have arisen in modern plasma physics. They are developing at an ever quickening tempo and owe their origin to the launching of artificial satellites of the Earth and space probes that pass through the magnetosphere and interplanetary plasma in their motion. One of the directions is connected with the study of interaction between bodies and plasmas. The other is connected with the study of the oscillations and waves that arise in plasmas, in particular as a result of interaction between the plasma and streams of particles.

The launching of man-made objects into the magnetosphere and interplanetary space has created a situation that to a certain extent resembles the situation that arose in the mechanics of continuous media after the invention of aeroplanes. Just as the development of aviation made it necessary to study the aerodynamic flow of a compressible gas around bodies, artificial satellites in the magnetosphere and interplanetary space have made it necessary to study the kinetics of plasma flow around bodies. The phenomena that occur around an object moving in a plasma are not decisive for its motion, in contrast to the motion of an aeroplane, since the frictional forces exerted by the plasma on satellites and rockets are small. However, these phenomena are, first, of considerable independent interest. They are distinguished by a number of features that have general significance for plasma physics. Second, their study has important bearing on the setting up and correct interpretation of many experiments on satellites and space probes, which are used as laboratories for investigating the properties of the surrounding medium.

On the other hand, the possibility of making direct measurements, as in a laboratory, in the magnetosphere and interplanetary plasma has led to the direct study of the wave processes in this plasma. This is also of general significance for plasma physics. Under laboratory conditions, it is virtually impossible to create a plasma with such a wide range of physical parameters as is found in the natural plasma. As a result, some of the phenomena predicted theoretically cannot be investigated at all under laboratory conditions. Moreover, the analysis of the results of experiments made with space probes is an excellent and one of the most exact methods of plasma diagnosis, enabling one, when the results of measurements can be identified with theoretical data, to determine the basic quantities that characterize the state of the plasma. For example, one can determine the wave modes that are observed, the nature of their excitation, and so forth. If, in addition, the wave processes are studied by means of instruments set up on the moving bodies themselves (satellites, rockets), the waves and plasma oscillations observed in the meighborhood of these probes can be associated with effects that arise as a result of the interaction of the body itself with the plasma and with the influence of the inhomogeneous ionized cloud and electric field formed around the body. Oscillations of the plasma can also be excited by streams of particles (for example, electrons) ejected from the body or electromagnetic waves (radio waves) transmitted from the body.

Thus, these two branches of modern plasma physics have many connections. They are also related by the unity of the theoretical methods used to solve different problems. In the majority of cases, kinetic plasma theory is used for this purpose.

In the majority of questions studied we are dealing primarily with a very rarefied magnetized plasma. The mean free path of the particles is much greater than the characteristic dimensions of the bodies and is frequently much greater than the wavelengths of the observed oscillations. The corresponding equations are written down in the phase space of the particles. The problems that have to be solved are very different from and more complicated than the problems of hydrodynamics. A distinguishing feature is the need to take into account the influence of electric and magnetic fields. The theory contains three new parameters with the dimensions of a length: the Debye radius, D, and the Larmor radii, ρ_{Hi} and ρ_{He}, of the ions and electrons. The flow of plasma around the bodies and also the spectra and modes of the waves and oscillations to be expected in the plasma become even more varied in the regions where the plasma is nonisothermal ($T_e \neq T_i$). The diversity of the expected wave

processes is also increased by inhomogeneous formations
of different scales that are always observed in the plasma and which, in their turn, are frequently a consequence
of the waves excited in the plasma. Theoretical problems
relating to plasma flow around the bodies, instability
of plasmas, and the wave modes excited in the plasmas
are frequently essentially nonlinear. This stage of
development of these directions of the physics of the
ionosphere, magnetosphere, and interplanetary plasma
characterizes the present state of the theory. The majority of effects described by the linear theory has already been fairly well studied from the theoretical side,
and also experimentally in many respects. This is undoubtedly a great achievement of this new field of experimental physics, It has been achieved in a compara-
tively short period in the last decade. However, the
further development of the theory requires the solution
of nonlinear problems. Here, each step forward requires
considerable exertions. Although modern computers can
now be used to solve some of the problems of this type,
there are as yet few results of the investigation of nonlinear problems. This is partly due to the fact that it
is difficult, from the many factors that influence the
investigated phenomena, to choose the principal one that
determines a given experimentally observed effect. This
has led to the accumulation of a number of experimental
data that have not yet received a clear and definite
theoretical explanation. This does not contradict the
assertion that the general state of these branches of
physics has reached a definitive form in a number of respects. Some of the results obtained experimentally and
theoretically have an elegant nature and illustrate the
richness of plasma physics and the possibilities it presents for studying nature and diagnosing the plasma surrounding the Earth. The aim of the present review is to
describe the principal results obtained in these fields.
I thought it would be a good idea to combine them in a
single book, since there is frequently an inner connection, as yet inadequately expressed, between the phenomena
that accompany the flow of plasma around bodies and the
wave processes observed in the plasma. In the following
sections, general equations are derived that describe
these phenomena, and in a number of cases the general
formulation of the corresponding problems is given. The
basic formulas that enable one to calculate the various
effects in the plasma are given, and the different types
of phenomenon are classified. As far as possible, experimental results are confronted with theoretical calculations. This reveals where theory and experiment have

been successfully united. It is for this reason that
the chapter on the flow of plasma around bodies even includes some results of laboratory experiments, since the
very approximate measurements in the near-Earth plasma
are as yet very sparse. It should be borne in mind that
the restricted length of the book -- brevity and compres-
sion of exposition were one of my aims -- have forced me
to omit a number of interesting (in particular, theoretical) results, and in some cases to give only a schematic exposition of the relevant data. However, the literature, which is cited in fair detail, will enable the
reader to become more acquainted with the questions in
which he is interested.

CHAPTER I

PROPERTIES AND PARAMETERS OF THE NEAR-EARTH AND INTER-PLANETARY PLASMA. BASIC EQUATIONS

§1. General Remarks

The phenomena that are considered in this book are observed at altitudes beginning at 200-300 km -- the region of maximal ionization in the ionopshere -- and extend to distances from the Earth of tens of thousands of kilometers. Use is also made of data of observations in the interplanetary medium and in the solar wind at distances from the Earth of $\sim 10^6$ km. The perigee of artificial satellites of the Earth is usally situated at the lower boundary of this region. It is here, approximately up to 1000-2000 km and somewhat higher, that the orbits of satellites most frequently pass and the majority of measurements is made. Satellite experiments are made less frequently at greater altitudes and in the transition region of the near-Earth plasma, i.e., in the region beginning at altitudes of 20-25 thousand kilometers, at which the upper boundary of the ionosphere is situated (this region is frequently called the plasmapause), up to distances of 10-15 Earth radii R_0, where the interplanetary medium proper begins. Even more infrequently are direct observations made in the interplanetary medium at great distances from the Earth; this occurs only when this region of space is traversed by space probes sent to the planets of the solar system or beyond its limits Thus, the information obtained about the properties of the natural plasma by direct measurements on bodies moving in the plasma is very varied for the different regions of the plasma. It is therefore natural that our current ideas concerning the phenomena in which we are here interested have been formed to a considerable extent by the results of the study of waves excited in the ionosphere and magnetosphere and detected on the surface of the Earth, and, in the case of plasma flow around bodies, by the results of a number of laboratory investigations.

The properties of this plasma vary in very wide ranges, and this naturally means that in a number of cases the physical phenomena occurring in the plasma have a very different nature; conversely, the same phenomena are observed both at the lower boundary of the plasma as well as in the magnetosphere. For example, this occurs when waves of the same mode but at very different frequencies are excited in the plasma. For this reason, it is important when considering the various phenomena, first, to divide the plasma into zones within which one can expect physical processes of the same type. For different phenomena the boundaries of these zones are of course different. Second, it is also important to split the whole of the very wide range of frequencies of the observed oscillatory phenomena into characteristic sections, though not in accordance with a quantitative principle, as has frequently been done hitherto in the literature, but rather on the basis of a definite physical approach to the wave processes. In this chapter, we outline more precisely an approach that can, from the physical point of view, be taken as the basis of such a classification and division of the plasma into zones. Before this, let us elucidate the problem with some examples.

When plasma flows around a body, the effects that arise in the neighborhood of the body are radically altered, this depending on the velocity V_0 of the body or the velocity v of the external streams of particles that encounter the body. The maximal velocities of the bodies launched into the plasma vary, as is well known, in the range 8-11 km/sec, i.e. $V_0 \sim 10^6$ cm/sec near the Earth, while at greater distances from the Earth the velocity is $V_0 \sim (2-5) \cdot 10^5$ cm/sec. In the considered regions of the plasma, the mean thermal velocity of the electrons varies in the range $v_e = \sqrt{2\kappa T/m} \approx 10^7 - 10^8$ cm/sec (see Tables 1 and 2 in §2), i.e., $v_e \gg V_0$. Therefore, with regard to the electrons, the body can always be assumed to be in a state of quasirest. However, with regard to the mean thermal velocity of the ions, which varies in the range $v_i = \sqrt{2\kappa T/M} \approx 10^5 - 10^6$ cm/sec and increases with increasing distance from the Earth, artificial satellites first move with supersonic velocity ($V_0 \gg v_i$), and then, in a certain intermediate region, $V_0 \sim v_i$, and finally there are then regions in which the body can be assumed to be in a state of quasirest ($V_0 < v_i$ or $V_0 \ll v_i$). Thus, in the different zones, the nature of the perturbation of the plasma around the body changes very strongly. We may point out that neglect of the simple circumstance that $V_0 \ll v_i$ at

great distances from the Earth meant that in some of the
early experiments with space probes the plasma densities
were determined incorrectly. We see that for the study
of questions of plasma flow around bodies it is sensible
to distinguish three regions:

zone I of s u p e r s o n i c   m o t i o n ($V_0 \gg v_i$), which
extends to altitudes 1000-2000 km;

zone II, the i n t e r m e d i a t e   z o n e, ($V_0 \sim v_i$),
encompasses the outer regions of the ionosphere: 2000
km < z < (3-5)R_0 (R_0 is the radius of the Earth);

zone III of q u a s i r e s t of the body ($V_0 \ll v_i$),
which encompasses primarily the interplanetary medium
(z > (10-15)R_0) and the solar wind.

With regard to the corpuscular streams -- the solar
wind -- that reach the Earth with a velocity V ~ 300-500
km/sec, the motion of satellites and space probes, like
that of the ions, can be regarded as supersonic in the
majority of cases.

However, the nature of the perturbation of the plasma in the neighborhood of moving bodies is determined
not only by V_0/v_i. Further important factors are the
linear size of the body, ρ_0, and the ratio of ρ_0 to the
Debye radius $D = (\kappa T/4\pi Ne^2)^{\frac{1}{2}}$ and the Larmor radii of the
ions and electrons, $\rho_{Hi} = v_i/\Omega_H$ and $\rho_{He} = v_e/\omega_H$ ($\Omega_H = eH_0/Mc$
and $\omega_H = eH_0/mc$ are the gyrofrequencies of the ions
and electrons and H_0 is the magnetic field of the Earth).
A l a r g e   b o d y is defined by the inequality $\rho_0 \gg D$.
The equations which must them be solved are the most
complicated, since one must take into account the boundary conditions on the surface of the body -- the properties of its surface (see §6). The phenomena around a
large body differ in a number of respects from those
around a s m a l l   b o d y, which one says has "point
size" when $\rho_0 \ll D$.   In the latter case, the problem can
be reduced to considering the motion of a point charge
in the plasma. In zone I, artificial bodies are basically large; in zone II one has initially $\rho_0 \sim D$. However, ρ_0 gradually becomes less than D, and in zone III
the bodies are small: $\rho_0 \ll D$.

The ratios ρ_0/ρ_{Hi} and ρ_0/ρ_{He}, which (especially the
first) affect the nature of the theoretical problems that
must be solved (and their difficulty) and the phenomena
observed in the neighborhood of the body, also vary within wide ranges in the different zones of the plasma. In
the majority of cases, $\rho_0 \ll \rho_{Hi}$ in all three zones. But
with regard to the precession of electrons we have $\rho_0 \gg$

ρ_{He}, $\rho_0 \ll \rho_{He}$, and $\rho_0 \ll \rho_{He}$ in zones I, II, and III, respectively. It can be seen from this treatment how varied are the limiting cases that must be solved in the different zones of the plasma. The hardest cases are those in which the characteristic parameters are commensurable, i.e., $V_0 \sim v_i$, $\rho_0 \sim D$, and $\rho_0 \sim \rho_{Hi}$.

With regard to the wave processes and resonances that occur in the ionosphere, magnetosphere, and interplanetary plasma, a division of the plasma into zones requires, in a number of respects, a different approach both when one considers the types of possible phenomena as well as the frequency ranges of the expected oscillations.

The conditions of excitation of waves in the plasma, the nature of its instability, and the spectra of its oscillations are of course strongly dependent on whether or not the plasma is strongly or weakly magnetized, i.e., on the ratio of the energy density $H_0^2/8\pi$ of the external magnetic field H_0 to the density $N\kappa(T_e + T_i)$ of the gas kinetic energy of the charged particles. This reduces to establishing which of the following conditions is satisfied:

$$(V_A/v_s)^2 \gg 1, \quad (V_A/v_s)^2 \sim 1, \quad (V_A/v_s)^2 \ll 1. \tag{1.1}$$

Depending on (1.1), the excited waves satisfy the conditions

$$\Lambda^2 = (v_{ph}/\omega)^2 \underset{\ll}{\overset{\gg}{\sim}} (\rho_{He})^2, \quad (\rho_{Hi})^2. \tag{1.2}$$

In formulas (1.1) and (1.2)

$$V_A = H_0/\sqrt{4\pi NM} = c\Omega_H/\Omega_0 \tag{1.3}$$

is the Alfvén velocity,

$$v_s = \sqrt{\kappa T_e/M} \tag{1.4}$$

is the velocity of nonisothermal sound, v_{ph} and Λ are the phase velocity and wavelength of the observed oscillations of the plasma, and $\Omega_0 = (4\pi Ne^2/M)^{\frac{1}{2}}$ is the ion plasma frequency. It is easy to see from the tables given in §2 that in all the regions in which we are interested the plasma is strongly magnetized, $V_A \gg v_s$, $\Lambda \gg \rho_{He}$, ρ_{Hi}. Therefore, a number of the wave phenomena in the near-Earth plasma has a universal character. These phenomena differ only in that they occur at frequencies that vary over several orders, this being due not to a difference

in the nature of the physical conditions in the plasma
but rather to changes in the values of the plasma parameters. For example, in different experiments one observes the excitation of ion cyclotron waves at low al-
titudes $z \sim 300{-}400$ km, where $\Omega_H/2\pi \sim 500{-}600$ Hz, and also
at distances from the Earth of $(25{-}30) \cdot 10^3$ km, where
$\Omega_H/2\pi \lesssim 1$ Hz. Or, for example, one detects the excitation of electron Langmuir oscillations (electron plasma
oscillations) at $z \sim 1000$ km, where $\omega_0/2\pi = (Ne^2/\pi m)^{\frac{1}{2}} \sim (2{-}3)$
$\cdot 10^6$ Hz and at a distance from the Earth of 10^6 km in the
solar wind, where $\omega_0/2\pi \sim (1{-}2) \cdot 10^4$ Hz. The frequency of
the lower hybrid resonance varies even more strongly, by
a factor 10^4, as one traverses the near-Earth plasma.
Thus, when one considers the dependence of the wave processes on the frequencies, the correct approach is to
classify the processes according to the type of the physical phenomenon that gives rise to them, i.e., one should
take the corresponding boundaries to be the characteristic
frequencies of the different processes. Such a classification will be given below in the following sections.

§2. Basic Parameters of the Near-Earth and Interplanetary Plasma

A common feature of all the parameters that characterize the different considered regions of the plasma is
their great variability depending on the time and the coordinates in any fixed region of altitudes. An exception
is the magnetic field of the Earth and all quantities related solely to it, for which the relative changes are
much smaller right up to distances from the Earth of several tens of thousands of kilometers. For example, the
electron density N at $z \sim 300{-}400$ km can vary from day to
night and with changing latitude and longitude by a factor of 10 or more. The electron temperature T_e can also
vary by a factor 5-6. At the same altitude, the ratio
$\Delta H_0/H_0$ is only of order $10^{-3}{-}10^{-4}$. In the interplanetary
medium, the relative variations of the magnetic field are
of course much stronger, and the density N of charged
particles at large distances from the Earth (in the interplanetary medium) are more stable, True, large variations
in the density of particles and of the magnetic field are
observed in the solar wind. The value of N is very un-
stable in the transition region of the near-Earth plasma,
in the plasmapause, especially at altitudes $z \sim 18{-}25$
thousand kilometers, where N may vary by a factor 100 or
more from case to case. There are data, not yet adequately verified, which indicate that such a situation obtains
right out to $z \sim 50{-}60$ thousand kilometers.

For these reasons, a sufficiently accurate analysis of the various phenomena considered below with a view to establishing a closer correspondence between experiments and theory is only possible if in one and the same experiment one undertakes a fairly large set of simultaneous measurements of different quantities, including a determination of the basic plasma parameters. Modern experiments on artificial satellites of the Earth or space probes enable one in principle to carry out such programs of investigation. In a number of cases, such experiments have been made, and some of the data obtained in them will be considered below. At the same time, comprehensive experiments of this kind are as yet rare, and in many cases estimates and calculations are made by means of averaged parameters of the plasma obtained under different conditions. In Tables 1 and 2 given here for the altitudes that correspond to the most characteristic zones of the near-Earth and interplanetary plasma (see §1) the corresponding values of the plasma parameters are close to their maximal values observed under different conditions.[14,15]

§3. Basic Equations and Properties of the Plasma

Usually, for the description of processes in a magnetized plasma one uses in the linear approximation the dispersion equation

$$A\tilde{n}^4 + B\tilde{n}^2 + C = 0, \tag{1.5}$$

where

$$\tilde{n} = n + i\kappa = c/v_{ph} + i\kappa \tag{1.6}$$

is the complex refractive index of the plasma, n is its real part, which determines the phase velocity v_{ph} of propagating waves, and κ is the spatial damping coefficient of such waves (Silin, Rukhadze;[1] Stix;[2] Akhiezer and others;[3] Ginzburg, Rukhadze[4]).

For the treatment of processes that occur in the plasma in time, it is more convenient to use a complex frequency:

$$\tilde{\omega} = \omega + i\gamma. \tag{1.7}$$

Then, depending on whether γ has a positive or negative value, the oscillations are damped or grow. With the description we choose of harmonic waves in the form exp $(i\tilde{\omega}t)$, γ is called the decay rate if $\gamma > 0$ and the

TABLE 1. Basic Parameters of the Near-Earth and Interplanetary Plasma

Zones	z	N, cm^{-3}	H_0, Oe	T_e^0	ν_{ei}, sec^{-1}	$N\kappa T$, erg/cm^3	$H_0^2/8\pi$, erg/cm^3	v_e, cm/sec	v_i, cm/sec	$\omega_H/2\pi$, sec^{-1}	$\Omega_H/2\pi$, sec^{-1}
Zone 1 $V_0 \gg v_i$ $H_0^2/8\pi \gg N\kappa T$	300 km	10^6	$4.5 \cdot 10^{-1}$	$1.5 \cdot 10^3$	$3 \cdot 10^3$	$2 \cdot 10^{-7}$	$8 \cdot 10^{-3}$	$2 \cdot 10^7$	$\sim 10^5$	$1.2 \cdot 10^6$	~ 40
	500 km	$2 \cdot 10^5$	$3.7 \cdot 10^{-1}$	$2 \cdot 10^3$	$3 \cdot 10^2$	$5 \cdot 10^{-8}$	$5 \cdot 10^{-3}$	$2.5 \cdot 10^7$	$1.4 \cdot 10^5$	10^6	~ 40
	2000 km	$4 \cdot 10^4$	$2.2 \cdot 10^{-1}$	$3 \cdot 10^3$	~ 10	$2 \cdot 10^{-8}$	$2 \cdot 10^{-3}$	$3 \cdot 10^7$	$8 \cdot 10^5$	$6 \cdot 10^5$	330
Zone 2 $V_0 \sim v_i$ $H_0^2/8\pi \gg N\kappa T$	$\sim R_0$	$5 \cdot 10^3$	$\sim 10^{-1}$	$6 \cdot 10^3$	<1	$4 \cdot 10^{-9}$	$4 \cdot 10^{-4}$	$6 \cdot 10^7$	$\sim 10^6$	$3 \cdot 10^5$	$1.6 \cdot 10^2$
	$3 \cdot 5R_0$	$5\text{-}100$	$\sim 10^{-2}$	$6 \cdot 10^4$	$\ll 1$	10^{-9}	$4 \cdot 10^{-6}$	$1.3 \cdot 10^8$	$3 \cdot 10^6$	$3 \cdot 10^4$	$1.6 \cdot 10$
Zone 3 $V_0 \ll v_i$ $H_0^2/8\pi > N\kappa T \sim ?$	$10\text{-}15R_0$	$5\text{-}10$	$5 \cdot 10^{-4}$	$\sim 10^5$	$\lll 1$	10^{-10}	$\sim 10^{-8}$	$1.8 \cdot 10^8$	$4 \cdot 10^6$	$1.4 \cdot 10^3$	$\lesssim 1$
	Interpl medium	$1\text{-}5$	$5 \cdot 10^5$	$\sim 2 \cdot 10^5$	$\lll 1$	$8 \cdot 10^{-11}$	$\sim 10^{-10}$	$2.5 \cdot 10^8$	$5 \cdot 10^6$	$1.4 \cdot 10^2$	$\lesssim 0.1$
	Solar Wind	$5\text{-}70$	$(5\text{-}20) \cdot 10^{-5}$	$(1\text{-}2) \cdot 10^5$	—	—	—	—	—	$(1.4\text{-}6) \cdot 10^2$	$(0.7\text{-}3) 10^{-1}$

TABLE 2. Basic Parameters of the Near-Earth and Interplanetary Plasma

Zones	z, km	$\rho_{He} = 2\pi v_e/\omega_H$, cm	$\rho_{Hi} = 2\pi v_i/\Omega_H$, cm	$\omega_0/2\pi$, sec^{-1}	$\Omega_0/2\pi$, sec^{-1}	$\sqrt{\omega_H \Omega_H}/2\pi$, sec^{-1}	$2\pi D = v_e \cdot (\sqrt{2}\omega_0)^{-1}$, cm	$n_A = \Omega_0/\Omega$	$v_A = c/n_A$, cm/sec
Zone 1 $\rho_0 \gg 2\pi D$	300 km	$1.5\cdot10$	$2.5\cdot10^3$	$9\cdot10^6$	$5\cdot10^4$	$7\cdot10^3$	1	$2.2\cdot10^3$	$2.5\cdot10^7$
$\rho_0 \ll \rho_{Hi}$	500 km	$2.5\cdot10$	$3.5\cdot10^3$	$4\cdot10^6$	$2\cdot10^4$	$6\cdot10^3$	4	$5\cdot10^2$	$6\cdot10^7$
$\rho_0 \gg \rho_{He}$	2000 km	$5\cdot10$	$2.5\cdot10^3$	$2\cdot10^6$	$5\cdot10^4$	$1.4\cdot10^4$	10	$1.5\cdot10^2$	$2\cdot10^8$
Zone 2 $\rho_0 \ll 2\pi D$	~R_0	$2\cdot10^2$	$6\cdot10^3$	$6\cdot10^5$	$1.5\cdot10^4$	$7\cdot10^3$	$7\cdot10^1$	10^2	$3\cdot10^8$
$\rho_0 \ll \rho_{Hi}$	$3.5R_0$	$4\cdot10^3$	$2\cdot10^5$	$5\cdot10^4$	10^3	$7\cdot10^2$	$2\cdot10^3$	$6\cdot10^2$	$5\cdot10^7$
$\rho_0 \ll \rho_{He}$									
Zone 3 $\rho_0 \ll 2\pi D$	(10–15)R_0 Interpl. medium	~10^5	~$4\cdot10^6$	$\geq3\cdot10^4$?	$2\cdot10^2$	$<4\cdot10$?	$4\cdot10^3$	$\leq2\cdot10^2$?	$\leq1.5\cdot10^8$?
$\rho_0 \ll \rho_{Hi}$	Solar	$2\cdot10^6$	$5\cdot10^5$	$>2\cdot10^4$?	~10^2	<4?	~10^4	$<10^3$?	$<3\cdot10^7$?
$\rho_0 \ll \rho_{He}$	Wind	–	–	$(2-7)10^4$	$(4.7–14)\cdot10^2$	–	–	–	–

growth rate if $\gamma < 0$. There is a simple relationship
between κ and γ:

$$\gamma = (\omega\kappa/c)\frac{d\omega}{dk}, \tag{1.8}$$

where $k = \omega n/c$ is the real part of the wave number.

The coefficients A, B, and C in the dispersion
equation (1.5) depend in the general case on the components of the 3×3 permittivity tensor,

$$\varepsilon_{ij}, \tag{1.9}$$

as follows:

$$A = \varepsilon_{11}\sin^2\theta + 2\varepsilon_{13}\sin\theta\cos\theta + \varepsilon_{33}\cos^2\theta,$$

$$B = -[\varepsilon_{11}\varepsilon_{33} + (\varepsilon_{22}\varepsilon_{33} + \varepsilon_{23}^2)\cos^2\theta - \varepsilon_{13}^2$$
$$+ (\varepsilon_{11}\varepsilon_{22} + \varepsilon_{12}^2)\sin^2\theta - 2(\varepsilon_{12}\varepsilon_{23} - \varepsilon_{13}\varepsilon_{22})\cos\theta\sin\theta], \tag{1.10}$$

$$C = \varepsilon_{33}(\varepsilon_{11}\varepsilon_{22} + \varepsilon_{12}^2) + \varepsilon_{11}\varepsilon_{23}^2 + 2\varepsilon_{12}\varepsilon_{13}\varepsilon_{23} - \varepsilon_{22}\varepsilon_{13}^2,$$

where θ is the angle between the wave vector $\mathbf{k}$ and the
vector H_0 of the external magnetic field. The components ε_{ij} themselves are determined from a selfconsistent solution of the kinetic equations with Maxwell's
equations for given conditions of the problem. For a
plasma consisting of two species of particles (electrons,
subscript e, and one species of ions, subscript i), the
corresponding system of kinetic equations in the nonstationary case has the form

$$\partial f_i/\partial t + V(\partial f_i/\partial r) + (eE/M)(\partial f_i/\partial v)$$
$$+ (e/Mc)[v + V, H_0](\partial f_i/\partial v) = 0, \tag{1.11}$$

$$\partial f_e/\partial t + V(\partial f_e/\partial r) - (eE/m)(\partial f_e/\partial v)$$
$$- (e/mc)[v + V, H_0](\partial f_e/\partial v) = 0.$$

For a moving body, the system (1.11) is solved simultaneously with the Poisson equation

$$\Delta\phi = -4\pi e[\int f_i d\overset{3}{v} - \int f_e d\overset{3}{v}],$$

$$E = -\text{grad } \phi. \qquad\qquad\qquad (1.12)$$

In equations (1.11) and (1.12), t is the time, r is the
vector that determines the position of the particle, v
is the velocity vector, ϕ and E are the potential and
the electric field, and $f_e(r,v,t)$ and $f_i(r,v,t)$ are
the distribution functions of the electrons and ions,
which in the general case depend on the spatial coordinates, the velocity, and the time. It is assumed
that the plasma has an ordered velocity V with respect
to the point of observation. In the case when one considers the problem of the motion of a body in a plasma
at rest, $V = V_0$, where V_0 is the velocity of the body.
If the ordered velocity V refers to a stream of particles
that encounters the plasma, terms must be added to the
corresponding equations to describe the distribution
functions of the particle streams, which are external
sources that act on the plasma.

Depending on the actual conditions of a problem,
one obtains different forms of the tensor elements,
which in their turn determine the nature of the phenomena observed in the plasma, in particular, the spectrum of the oscillations of the plasma. It is natural
that specific problems require a formulation of a definite type of conditions of the plasma state; for example, the temperature of the plasma particles, the
boundary conditions (on the surface of the moving bodies),
the nature of the sources (external electric fields,
incident waves, particle streams). In addition, as, for
example, in the problems of flow around a body (Al'pert,
Gurevich, Pitaevskii[5]), one must also write down on the
right side of (1.11) the collision integrals Y_e and Y_i,
which take into account the influence of collisions between particles on the distribution function. It is
true that in a number of cases one can ignore the influence of the collision integral. However, for example,
for the study of the scattering of radio waves on the
wake of a body, it is of fundamental importance to allow
for collisions, since this restricts the divergence of
the formulas that are obtained (see ref. 5). It is the
nature of these conditions that determines the type of
phenomena observed. The corresponding concrete cases
that have been investigated experimentally and theoretically are discussed in the following sections; this
is indeed the subject and aim of the book. Here, however, we continue our study of the general properties
of the plasma and its parameters.

In unperturbed regions of the plasma, the distribution functions f_e and f_i are Maxwellian and depend only on the particle velocities v and, for example, at sufficiently great distances from the moving body, where the state of the plasma is only very slightly disturbed,

$$f_{i0} = N_{0i} (M/2\pi\kappa T)^{3/2} \exp [-M_i (v + V_0)^2/2\kappa T],$$

$$\text{(1.13)}$$

$$f_{e0} = N_{0e} (m/2\pi\kappa T)^{3/2} \exp [-m(v + V_0)^2/2\kappa T],$$

and N_{0e} and N_{0i} are the unperturbed electron and ion densities. In the perturbed regions, the corresponding particle densities are determined by the integrals

$$N_i (r, t) = \int f_i (r, v, t) d\overset{3}{v},$$

$$\text{(1.14)}$$

$$N_e (r, t) = \int f_e (r, v, t) d\overset{3}{v}.$$

Note that if problems of plasma flow around bodies are solved in a coordinate system attached to the moving bodies, the time dependence of all quantities disappears ($\partial f/\partial t = 0$), and the problem becomes stationary. Naturally, in the absence of an influence of ordered motion, $V_0 = 0$ in (1.13). When one considers the effect of streams of particles on the plasma (beams of electrons or ions) with, for example, Maxwellian distributions, it is necessary in (1.13) for the beam to replace N_{0e} or N_{0i} by the particle densities of the beams and V_0 by the velocity V of the beam. We mention here that a Maxwellian distribution is distinguished by the fact that $\partial f_0/\partial v = 0$ for $v = 0$ if $V_0 = 0$, while $\partial f_0/\partial v < 0$ when $v \neq 0$ in the whole range of velocities ($v > 0$). This means in an equilibrium plasma one always has $\gamma > 0$ (see (1.7)), i.e., the oscillations of the plasma are damped and processes of growth of oscillations or the excitation of waves cannot occur. It was shown in ref. 6 (Landau) that in the general case for an arbitrary distribution function

$$\gamma \sim -\partial f/\partial v. \qquad\qquad \text{(1.15)}$$

Therefore $\gamma > 0$ if $\partial f/\partial v < 0$. However, one can have distribution functions for which $\partial f/\partial v > 0$ in a certain range of velocities. For example, this can occur if there is in the plasma a beam of particles whose ordered velocity exceeds the equilibrium thermal velocities of the plasma in a certain interval of velocities. In this region of velocities, the total distribution function has an increasing branch with positive derivative, $\partial f/\partial v > 0$, and here $\gamma < 0$. This results in the possible growth of oscillations of the plasma and excitation of waves in the plasma. Essentially, this is the t w o - s t r e a m   i n s t a b i l i t y of a plasma -- the possibility of the growth of resonance oscillations in the plasma.

We point out here an important circumstance that greatly simplifies the solution of a certain type of problem in the near-Earth and interplanetary plasma; for example, problems relating to the plasma flow around artificial bodies that move in the plasma with velocities satisfying $V_0 \ll v_e$. This last condition enables one to assume that the electrons have a Maxwell-Boltzmann distribution, i.e.,

$$f_e = f_e(\mathbf{v}) = N_{0e}(m/2\pi\kappa T)^{3/2}$$
$$\times \exp [e\phi/\kappa T - m(\mathbf{v} + \mathbf{V_0})^2/\kappa T].$$

(1.16)

Therefore, the second equation in the system (1.11) vanishes, and the Poisson equation (1.12) simplifies and can be rewritten in the form

$$\Delta\phi = -4\pi e\{\int f_i d\mathbf{v}^3 - N_{0e}\exp(e\phi/\kappa T)\}.$$

(1.17)

Naturally, as in the system of equations (1.11) and (1.12), the potential in (1.17) depends on $\mathbf{r}$ and t, $\phi = \phi(\mathbf{r},t)$, if one considers a nonstationary problem, while $\phi = \phi(\mathbf{r})$ if one is solving a stationary problem, i.e., if it is assumed in (1.11) that $\partial f/\partial t = 0$.

The plasma properties when they are described on the basis of kinetic theory, which is dictated here, for example, by the fact that the critical phenomena in which we are interested depend on both the disordered and the ordered velocities of the particles of different species, are characterized, as is well known, by spatial and frequency dispersion. F r e q u e n c y   d i s p e r s i o n is manifested in the fact that the different critical quantities in the considered phenomena depend on the frequency f ($\omega = 2\pi f$ is the angular frequency). This means that the state of the plasma at a given time depends on processes that have taken place in the preceding time. This is how the t e m p o r a l   i n e r t i a of the plasma is manifested. S p a t i a l   d i s p e r s i o n is manifested in the fact that the different parameters depend on the wave vector $\mathbf{k}$. This means that the state of a plasma at a given point depends on phenomena occurring in the region surrounding the given point -- in principle, at any point of the plasma. This is the s p a t i a l i n e r-t i a of the plasma, associated with the transfer of "influence" from one point to another. In particular, the elements ε_{ij} of the permittivity tensor are functions of ω and $\mathbf{k}$. Usually, one also introduces the complex conductivity tensor, which is related to ε_{ij} by

$$\varepsilon_{ij}(\omega,\mathbf{k}) = \delta_{ij} + (4\pi i/\omega)\sigma_{ij}(\omega,\mathbf{k}), \tag{1.18}$$

where σ_{ij} are the elements of the complex conductivity tensor of the plasma, which depends on the polarizability of the different species of particles and is determined from the solution of the system of equations (1.11) and (1.12), and $\delta_{ij} = 0$ for $i \neq j$ and $\delta_{ij} = 1$ for $i = j$.

Similarly, the complex refractive index (1.6) and the complex frequency (1.7) should be written in the form

$$\tilde{n}(\omega,\mathbf{k}) = n(\omega,\mathbf{k}) + i\kappa(\omega,\mathbf{k}),$$

$$\tag{1.19}$$

$$\tilde{\omega}(\mathbf{k}) = \omega(\mathbf{k}) + i\gamma(\mathbf{k}).$$

Therefore, the dispersion equations are frequently written not in the form (1.5) but in the form

$$F(\omega,\mathbf{k}) = 0 \quad \text{or} \quad \omega = \omega(\mathbf{k}), \tag{1.20}$$

which is not only a more convenient form of expression, but sometimes also enables one to penetrate more readily into the essence of the phenomena.

It is important to point out a further fundamental general property of the permittivity which, in particular, has great importance for the questions in which we are interested. The integrands that determine the permittivity tensor always have singular points, these being determined in the general case by the conditions of resonance in the plasma -- the conditions under which the plasma particles interact most strongly with the field of the waves. These conditions arise when a field interacts with either the electrons or the ions of the plasma, and they have the form

$$\omega = kv_{\|},$$
$$\omega = kv_{\|} + s\omega_H, \tag{1.21}$$
$$\omega = kv_{\|} - s\Omega_H,$$

where $kv_\parallel = k \cdot v_\parallel \cos \theta$, in which θ is the angle between k and H_0, $s = \pm(1,2,3,\ldots)$, and $v_\parallel$ is the longitudinal (along the vector H_0) component of the mean velocity of the particles. Note that for particle velocities near c it is necessary to take into account relativistic effects in (1.21), and ω_H and Ω_H are replaced by $\omega_H(1 - v^2/c^2)^{1/2}$ and $\Omega_H(1 - v^2/c^2)^{1/2}$, where v is the total velocity of the particle.

The first of the conditions (1.21) describes the C h e r e n k o v - V a v i l o v e f f e c t and determines the conditions of so-called Cherenkov damping or, conversely, excitation of oscillations of the plasma. If the phase velocity ω/k of the waves is greater than the longitudinal velocity component $v_\parallel$ of the particles, the oscillations are damped since the particles obtain from the field more energy than they give up. The opposite phenomenon, so-called Cherenkov excitation, occurs when $\omega/k < v_\parallel$, when the particles interacting with the waves take up from the latter less energy than they give up. At the same time, it is clear that the condition $\omega = kv_\parallel \cos \theta$ can be satisfied only if $\cos \theta > 0$. This means that the Cherenkov radiation occurs in the same direction in which the particle moves.

The other two conditions (1.21) when $s \neq 0$ describe c y c l o t r o n e x c i t a t i o n or damping of waves, this also depending on whether the phase velocity of the wave is greater or less than the particle velocity. The physical meaning of the terms $k \cdot v_\parallel$ in (1.21) is that they determine the Doppler shifts of the frequencies of the excited oscillations. When $s > 0$, the Doppler effect is normal -- in this case the phase velocity v_{ph} of the wave is greater than $v_\parallel$. When $s < 0$, we have the anomalous Doppler effect: $v_{ph} < v_\parallel$. It is readily seen that, depending on the sign s, the condition of gyroresonance is satisfied for $\cos \theta > 0$ or $\cos \theta < 0$. In the first case $\theta < \pi/2$, and the radiation of the particle is directed in the direction of its motion, which corresponds to the anomalous Doppler effect, as in the case of Cherenkov radiation. In the second case $\theta > \pi/2$ the direction of the radiation is opposite to that of the motion of the particle; this corresponds to the normal Doppler effect.

A very important property of a h o t (k i n e t i c)

plasma is that even under conditions when one can ignore collisions between its particles, i.e., in a collision-less plasma (the collision frequencies vanish, $\nu = 0$), the oscillatory processes are damped because of their interaction with particles ($\gamma < 0$). The absence of this damping, the so-called L a n d a u d a m p i n g (Cherenkov or gyroresonance damping) is one of the main qualitative differences of a c o l d plasma, i.e., a plasma in which the influence of the thermal motion of the particles can be ignored. As is well known, in contrast to col-lisionless damping, resonant phenomena also occur in a cold plasma, for which the theory is constructed on the basis of the equations of hydrodynamics (or quasihydro-dynamics) or on the basis of the microfield equations.

§4. Refractive Indices and Resonances of a Cold Plasma (T = 0). Classification of Waves

If one ignores the influence of thermal motion of the particles and there are no particle streams, the permittivity tensor takes its simplest form. The ele-ments of the tensor depend in this case solely on the frequency (there is no spatial dispersion) and for a multicomponent plasma (consisting of electrons and sev-eral species of ions) without allowance for collisions between the electrons and ions, they have the form

$$
\begin{aligned}
\varepsilon_{11} &= \varepsilon_{22} = \varepsilon_1 = 1 - \omega_0^2/A - \Omega_{01}^2/B_1 - \Omega_{02}^2/B_2 - \ldots, \\
\varepsilon_{12} &= i\varepsilon_2 = i\omega_0^2\omega_H/\omega A - i\Omega_{01}^2\Omega_{H1}/\omega B_1 \\
&\qquad - i\Omega_{02}^2\Omega_{H2}/\omega B_2 - \ldots, \\
\varepsilon_{33} &= \varepsilon_3 = 1 - \omega_0^2/\omega^2 - \Omega_{01}^2/\omega^2 - \Omega_{02}^2/\omega^2 - \ldots, \\
A &= \omega^2 - \omega_H^2, \quad B_i = \omega^2 - \omega_{Hi}^2, \quad i = 1, \ 2,
\end{aligned}
\tag{1.22}
$$

where the subscripts 1, 2,... on the right sides of (1.22) refer to different species of ions. When allow-ance is made for collisions between only electrons and ions in the expressions for the plasma frequencies and gyrofrequencies,

$$
\omega_0^2 = 4\pi Ne^2/m, \quad \Omega_{01}^2 = 4\pi Ne^2/M_1, \quad \ldots
$$

$$
\tag{1.23}
$$

$$
\omega_H = eH_0/mc, \quad \Omega_{H1} = eH_0/M_1c, \quad \ldots
$$

it is necessary to replace the particle masses by the
values

$$m(1 + i\nu_{ei}/\omega), \quad M_1(1 + i\nu_{ii}/\omega), \quad \ldots \qquad (1.24)$$

where ν_{ei} are the collision frequencies for collisions
between electrons and the different species of ions and
$\nu_{1i}, \nu_{2i}, \ldots$ are the collision frequencies for collisions
between ions of different species. It follows from the
dispersion equation (1.5) that

$$n_{12}^2 = \frac{-B \pm (B^2 - 4AC)^{\frac{1}{2}}}{2A} \qquad (1.25)$$

where for a cold plasma

$$A = \varepsilon_1 \sin^2\theta + \varepsilon_3 \cos^2\theta,$$
$$B = -\varepsilon_1\varepsilon_3(1 + \cos^2\theta) - (\varepsilon_1^2 - \varepsilon_2^2)\sin^2\theta, \qquad (1.26)$$
$$C = \varepsilon_3(\varepsilon_1^2 - \varepsilon_2^2).$$

We mention here that in the greater part of the plasma
with which we are here concerned the phenomena are described by the formulas for a two-component plasma (consisting of electrons and protons). Therefore, except
for the cases when the multicomponent nature of the plasma
plays a role, such effects are described below, and we
give here primarily the formulas for a two-component
plasma. Corresponding compact formulas, for example,
for n^2, can be obtained in the general case for every
range of frequencies in two limiting cases: when $\theta = 0$

$$(n^2 - \kappa^2)_{12} = 1 - \frac{\omega_0^2(\omega \pm \omega_H)(\omega \mp \Omega_H)}{(\omega \pm \omega_H)^2(\omega \mp \Omega_H)^2 + \nu^2\omega^2}, \qquad (1.27)$$

$$(2n\kappa)_{12} = \frac{\omega_0^2 \cdot \nu\omega}{(\omega \pm \omega_H)^2(\omega \mp \Omega_H)^2 + \nu^2\omega^2},$$

and when $\theta = \pi/2$

$$\tilde{n}_1^2 = 1 - \frac{\omega_0^2}{\omega - i\nu\omega}, \qquad (1.28)$$

$$\tilde{n}_2^2 = 1 - \frac{\omega_0^2}{\omega^2 - \Omega_H\omega_H - i\nu\omega - \omega^2\omega_H^2[\omega^2 - \omega_0^2 - \Omega_H\omega_H - i\nu\omega]^{-1}}$$

In (1.27) and (1.28) we have also allowed for collisions between electrons and neutral particles, i.e., in these equations $\nu = \nu_{ei} + \nu_{en}$.

We should also derive compact formulas that are obtained for the case of quasilongitudinal propagation, which is of interest when one considers the guiding of electromagnetic waves in the neighborhood of a line of force of the geomagnetic field, or when one considers the capture of waves in the so-called magnetic ducts. For electron low-frequency waves (see below), i.e., for frequencies

$$\omega_H \gtrsim \omega > \omega_L$$

(ω_H is the lower hybrid frequency; see (1.29)), the condition of quasilongitudinal propagation is

$$\frac{\sin^2\theta}{2\cos\theta} \ll \left| \frac{\omega^2 - \omega_0^2 - i\nu\omega}{\omega\omega_H} \right|.$$

In this case,

$$\tilde{n}_{12}^2 \approx 1 - \frac{\omega_0^2}{\omega(\omega \pm \omega_H \cos\theta - i\nu)} \tag{1.27a}$$

At frequencies $\omega < \sqrt{\omega_H\Omega_H} \sim \omega_L$ corresponding to the ranges of very low frequency (VLF) and extra low frequency (ELF) waves (see below), the condition of quasilongitudinal propagation

$$\frac{\sin^2\theta}{2\cos\theta} \ll \frac{\omega\left|\Omega_0^2 + i\gamma_0\nu\omega\right|}{\Omega_H\left|\omega^2 - \Omega_0^2 - \Omega_H^2 + i\gamma_0\nu\omega\right|}$$

(where $\gamma_0 = m/M$) leads to

$$\tilde{n}_{12}^2 = 1 + A/(B + C),$$
$$A = \Omega_0^2(\Omega_0^2 + \Omega_H^2\sin^2\theta + i\gamma_0\nu\omega),$$
$$B = \Omega_0^2\Omega_H\cos\theta(\Omega_H\cos\theta \pm \omega - \gamma_0^2\nu^2\omega^2/\cos\theta), \tag{1.27b}$$
$$C = i\gamma_0\nu\omega(\Omega_0^2 + \Omega_H^2\sin^2\theta \pm \omega\Omega_H\cos\theta),$$

which under the conditions $\Omega_H^2 \ll \Omega_0^2$ and $\gamma_0\nu\omega \ll \Omega_0^2$, which are realized in the near-Earth plasma, takes on a compact form similar to (1.27a), namely

$$\tilde{n}_{12}^2 = 1 + \Omega_0^2/[\Omega_H \cos\theta\,(\Omega_H \cos\theta \pm \omega) + i\gamma_0\nu\omega]. \qquad (1.27c)$$

For arbitrary θ, i.e., when the condition of quasi-longitudinal propagation is not satisfied, equation (1.27c) is replaced by the following fairly simple formula for the refractive index:

$$\tilde{n}_{12}^2 = 2\Omega_0^2/[\Omega_H^2(1 + \cos^2\theta \mp A) + 2i\gamma_0\nu\omega],$$

$$(1.27d)$$

$$A = [\sin^4\theta + (4\omega^2/\Omega_H^2)\cos^2\theta]^{\frac{1}{2}}.$$

It follows from this in particular that in the neighborhood of the ion gyrofrequency, i.e., for $\omega \sim \Omega_H$,

$$\tilde{n}_1^2 = \frac{\Omega_0^2(1 + \cos^2\theta)}{2\Omega_H(\Omega_H - \omega)\cos^2\theta + i\gamma_0\nu\omega(1 + \cos^2\theta)},$$

$$(1.27e)$$

$$\tilde{n}_2^2 = \Omega_0^2/\{\Omega_H^2(1 + \cos^2\theta) + i\gamma_0\nu\omega\}.$$

and for $\omega/\Omega_H \ll 1\,(\theta > 0)$, when $\sin^4\theta/4\cos^2\theta \gg \omega^2/\Omega_H^2$

$$\tilde{n}_1^2 = \frac{\Omega_0^2}{\Omega_H^2\cos^2\theta + i\gamma_0\nu\omega},$$

$$(1.27f)$$

$$\tilde{n}_2^2 = \frac{\Omega_0^2}{(\Omega_H^2 + i\gamma_0\nu\omega)}.$$

In a collisionless plasma $(\nu = 0)$, equations (1.27) and (1.28) are replaced, respectively, for $\theta = 0$ by

$$n_{12}^2 = \frac{\omega^2 \mp \omega\omega_H - \Omega_H\omega_H - \omega_0^2}{(\omega \pm \omega_H)(\omega \mp \Omega_H)}, \qquad (1.27g)$$

and for $\theta = \pi/2$ by

$$n_2^2 = \frac{(\omega - \omega_-^2)(\omega^2 - \omega_+^2)}{(\omega - \omega_L^2)(\omega^2 - \omega_U^2)}, \qquad (1.28a)$$

where ω_- and ω_+ are the roots of the equation

$$\omega_\mp^2 \mp \omega_\mp\omega_H - \Omega_H\omega_H - \omega^2 = 0.$$

The numerator of (1.28a) determines the zeros of the refractive index, and ω_L and ω_U in the denominator of (1.28) are the so-called lower and upper hybrid frequencies (see below):

$$\omega_L^2 = \Omega_H \omega_H (1 + \omega_H^2/\omega_0^2)^{-1} \quad \text{or} \quad 1/\omega_L^2 = 1/\Omega_H \omega_H + 1/\Omega_0^2 ,$$

$$\omega_U^2 = \omega_H^2 + \omega_0^2 \tag{1.29}$$

It should be pointed out that formulas (1.22) and (1.26) have a larger region of applicability than for a cold plasma, when the temperatures of the particles are taken equal to zero: $T_e = T_i = 0$. Namely, they can be used when

$$k_\perp v_e^2/\omega_H = \rho_{He}/\Lambda_{\perp e}^2 \ll 1,$$

$$(k_\perp v_i/\Omega_H)^2 = (\rho_{Hi}/\Lambda_{\perp i})^2 \ll 1,$$

$$(\omega - s\omega_H)/k_{\parallel} v_e = (1 - s\Omega_H/\omega)(v_{ph\parallel e}/v_i) \gg 1,$$

where $s = 0, \pm 1, \pm 2$ and the subscripts $\perp$ and $\parallel$ denote the normal and longitudinal components of the wave vector k and the wavelength Λ along H_0. As we see, these conditions mean that the approximation of a cold plasma is good when the wavelengths of the low- and high-frequency oscillations are greater than the ion and electron Larmor radii, respectively, and the phase velocities of the waves are high compared with the thermal velocities of these particles. This approximation fails only in the regions sufficiently near the gyroresonances Ω_H, $2\Omega_H$, ω_H, and $2\omega_H$ (see ref. 3).

Since there is no spatial dispersion in a cold plasma and the coefficients A, B, and C do not depend on k, equation (1.25) is an algebraic equation of fourth order in n. It determines two values of $n_{12}^2(\omega,\theta)$, two modes of elliptically polarized waves -- ordinary (n_1) and extraordinary (n_2) -- with different phase velocities and polarization senses. The ordinary wave has left-handed rotation of the electric vector (i.e., counter clockwise) with respect to the direction of the wave vector, while the extraordinary wave has a right-handed rotation. Both waves are transverse. However, in a two-component cold plasma this equation has five branches of both modes of transverse wave -- two branches $n_1(\omega,\theta)$ and three branches $n_2(\omega,\theta)$. The corresponding dependences $n_{12}^2(\omega)$ and $\omega_{12}(k)$ are shown schematically in Figs. 1 and 2. Before we describe the general properties of these curves, which will enable us to identify them with the terminology adopted in the literature (the names of the different types of wave) and introduce a classification of them (division into frequency ranges) let us dwell on the following very important property of the dispersion equation (1.25). It can be seen that $n^2 \to \infty$ when A = 0. In this case

$$n_1^2 = C/B \quad \text{and} \quad n_2^2 = -B/A.$$

Since $B \neq 0$, the refractive index can diverge for only one of the waves, namely, only $n_2^2 \to \infty$. The condition $A \to 0$ leads to the equation

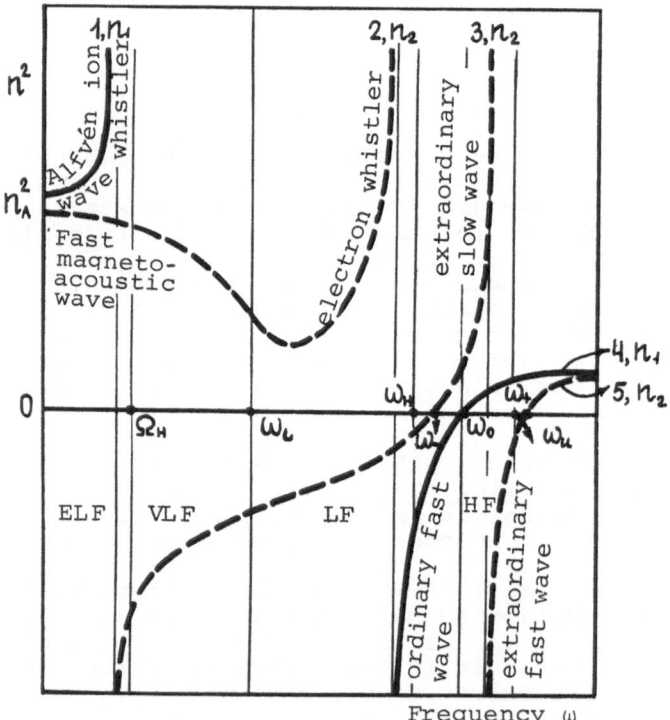

Fig. 1

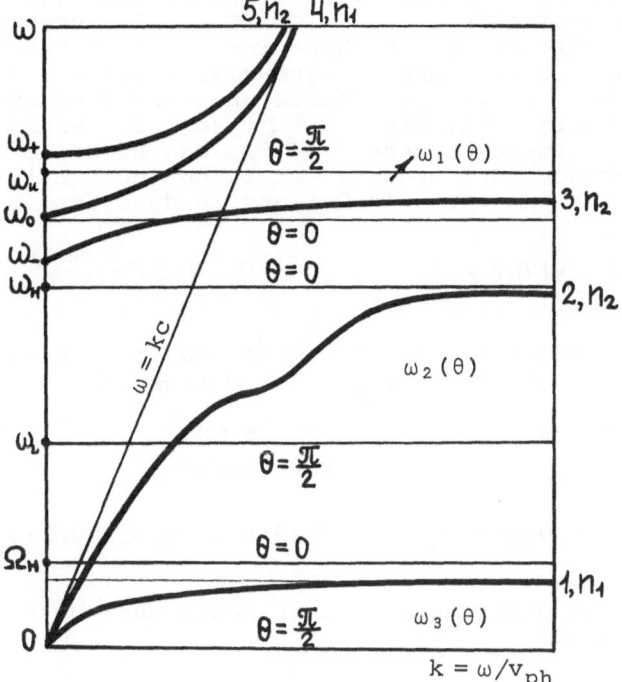

Fig. 2

$$1 - \frac{\omega_0^2 \cos^2\theta}{\omega^2} - \frac{\omega_0^2 \sin^2\theta}{\omega^2 - \omega_H^2} - \frac{\Omega_0^2 \cos^2\theta}{\omega^2} - \frac{\Omega_0^2 \sin^2\theta}{\omega^2 - \Omega_H^2} = 0 \qquad (1.30)$$

This is an equation of third degree in ω^2 and it determines three resonance branches $\omega(\theta)$ under which characteristic longitudinal oscillations directed along the wave vector can arise in a cold plasma. Since there is no spatial dispersion in the considered approximation (the group velocity vanishes, $d\omega/dk = 0$), one cannot, strictly, speak in this approximation of the resonant excitation of longitudinal waves; for the oscillations cannot escape from the region in which they are excited.

These resonance branches $\omega_1(\theta)$, $\omega_2(\theta)$, and $\omega_3(\theta)$ are
shown schematically in Fig. 3. (Spatial dispersion has
a strong influence on the behavior of n_2^2 in the neighborhood of the resonance frequencies $\omega_{1,2,3}(\theta)$. However,
their values are changed little and they have only additional factors $1 + \xi$, where $\xi \ll 1$; see, for example, ref.
3.) One of these branches, $\omega_1(\theta)$, can be called the
high-frequency (HF) branch. It is not much affected by
the ions (ω_0, $\omega_H \gg \Omega_0$ and Ω) and is described by

$$\omega_1^2(\theta) = \tfrac{1}{2}\left\{ (\omega_0^2 + \omega_H^2) + \left[(\omega_0^2 + \omega_H^2)^2 - 4\omega_0^2\omega_H^2\cos^2\theta \right]^{\tfrac{1}{2}} \right\}.$$

$$(1.31)$$

If $\omega_0 > \omega_H$ (which is almost always realized in our case;
see Tables 1 and 2 in §2), $\omega_1(\theta)$ varies in the ranges

$$\omega_1(\theta) = \omega_0 \qquad\qquad \text{for} \quad \theta = 0,$$
$$\omega_2(\theta) = \omega_U = (\omega_0^2 + \omega_H^2)^{\tfrac{1}{2}} \quad \text{for} \quad \theta = \pi/2. \qquad (1.32)$$

The frequency ω_U is called the upper hybrid frequency.

The two other resonance branches must be determined
with allowance for the ions:

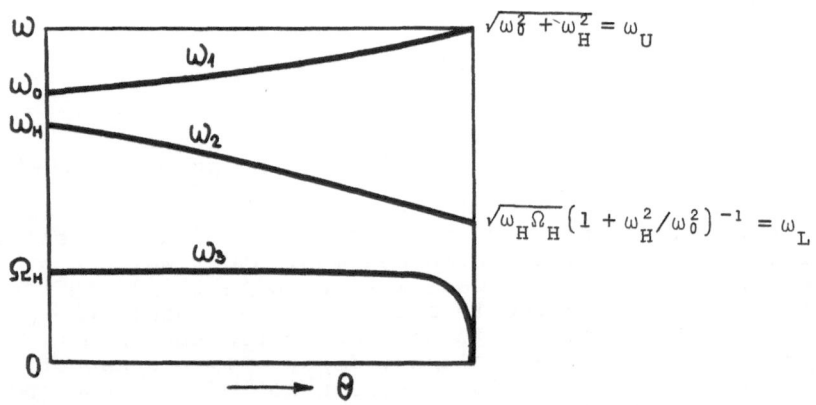

Fig. 3

$$\omega_{23}(\theta) = \tfrac{1}{2}\omega_H^2(\omega_0^2 + \omega_H^2)^{-1}(\omega_0^2\cos^2\theta + \Omega_0^2 + \Omega_H^2 \pm A),$$

$$A = \{(\omega_0^2\cos^2\theta + \Omega_0^2 + \Omega_H^2)^2 - 4(\omega_0^2 + \omega_H^2)\omega_H^{-2}(\Omega_H^2\omega_0^2\cos^2\theta)\}^{\tfrac{1}{2}}.$$

$$(1.33)$$

We call the branch $\omega_2(\theta)$ the low-frequency (LF) branch; it varies in the ranges

$$\omega_2(\theta) = \omega_H \qquad\qquad\qquad\quad \text{for} \quad \theta = 0$$
$$\omega_2(\theta) = \omega_L; \quad 1/\omega_L^2 = 1/\omega_H\Omega_H + 1/\Omega_0^2 \quad \text{for} \quad \theta = \pi/2.$$

$$(1.34)$$

The frequeny ω_L is called the lower hybrid frequency.

The third branch is the extra low frequency (ELF) branch; it varies in the ranges

$$\omega_3(\theta) = \Omega_H \quad \text{for} \quad \theta = 0,$$
$$\omega_3(\theta) \approx \Omega_H \cos\theta[(\Omega_H^2 + \Omega_0^2)/\omega_0^2]^{-\tfrac{1}{2}} = 0 \quad \text{for} \quad \theta = \pi/2.$$

$$(1.35)$$

A very important property follows from these results: in a cold plasma there are no resonant oscillations in two frequency sections: first, between the ion gyrofrequency Ω_H and the lower hybrid frequency ω_L. In this section of frequencies, as we shall see below, oscillations are excited in a nonisothermal plasma, when allowance must be made for the thermal motion of the particles, if $T_e \gg T_i$. It is expedient to call the oscillations in this frequency interval very low frequency (VLF) oscillations. Note that in the VLF range of frequencies, i.e., for $\Omega_H < \omega \lesssim \omega_L$, the influence of ions of different species (see (1.43)) gives rise in the near-Earth plasma to some very interesting effects: the capture of waves, wave cutoff, and complicated types of wave path. These effects are considered in Chapter III.

Second, in a cold plasma there are no resonances in the section that adjoins the high frequency oscillations, between ω_H and ω_0, where resonances are also possible only when spatial dispersion is taken into account.

We now describe briefly the five branches $n^2(\omega,\theta)$ and $\omega(k,\theta)$ shown in Figs. 1 and 2, that is, the general properties of the five waves that can propagate in a cold plasma. Below and in Chapter III we also deduce some formulas with allowance for spatial dispersion.

The branches 1 in Figs. 1 and 2 correspond to the root n_1^2 of equation (1.25). This is an ordinary ion wave. It is cut off at the ion gyrofrequency. As $\omega \rightarrow \Omega_H$, its refractive index tends to infinity, $n_1^2 \rightarrow \infty$. For $\omega > \Omega_H$, this wave is strongly damped and cannot propagate. The cutoff of the wave is of a purely "kinetic" origin and is associated with the ion-cyclotron resonance $\omega \rightarrow \Omega_H$. This wave is sometimes called an ion whistler (see Chapter III). As $\omega \rightarrow 0$ ($\omega \ll \Omega$) the branch 1 describes the properties of the so-called Alfvén wave:

$$\omega = kV_A \cos \theta [1 + (V_A/c)^2]^{-\frac{1}{2}},$$

$$n_{10} = n_A [1 - (\omega/\Omega_H) \cos \theta]^{-1},$$

$$\kappa_{10} = \tfrac{1}{2} n_{10} \nu \omega [\omega_H (\Omega_H - \omega)]^{-1}, \qquad (1.36)$$

where $\quad n_A = \Omega_0/\Omega_H \quad$ and $\quad V_A = c/n_A$.

In (1.36), n is the Alfvén refractive index (see (1.13)) and κ_{10} is the collisional damping coefficient of the wave. Here and in all that follows, the subscript 0 appended to different quantities signifies their value in a cold plasma. For the cyclotron wave ($\omega \rightarrow \Omega_H$) we have

$$n_{10}^2 \approx n_A^2 (1 + \cos^2\theta) [(1 - \omega/\Omega_H) 2\cos^2\theta]^{-1}. \qquad (1.36a)$$

The general formula for n_{12}^2 in the considered range of frequencies up to $\omega \gtrsim \omega_L$ (see (1.29)) has the form

$$(n_{12}^2)_0 = (2\cos^2\theta)^{-1} [\varepsilon_1 (1 + \cos^2\theta) \pm A],$$

$$A = [\varepsilon_1^2 (1 + \cos^2\theta - 4(\varepsilon_1^2 - \varepsilon_2^2) \cos^2\theta]^{\frac{1}{2}}, \qquad (1.37)$$

where

$$\varepsilon_1 = \Omega_0^2 (\omega^2 - \omega_H^2)^{-1},$$

$$\varepsilon_2 = -\omega\Omega_0^2 [\Omega_H (\omega^2 - \Omega_H^2)]^{-1}. \qquad (1.38)$$

In the general case, the dispersion equation is more complicated, so we omit it. For $\theta = 0$, however, it has a simple form and for a frequency range that also encompasses the frequencies of branch 2, i.e., as $\omega \rightarrow \omega_H$

$$F(\omega,k)_{12} = c^2k^2 - \omega^2 + \omega_0^2\omega[\omega \pm \omega_H]^{-1}$$
$$+ \Omega_0^2\omega[\omega \mp \Omega_H]^{-1}. \tag{1.39}$$

The frequency range encompassed by branch 1 corresponds to the region of frequencies of the ELF waves $\omega_3(\theta)$ described by equations (1.33) and (1.35) (see Fig. 3).

Branch 2 in Figs. 1 and 2 describe waves that cover the frequency range $0 \leqslant \omega \leqslant \omega_H$. These are electron extraordinary waves n_2. They are cut off by the cyclotron resonance of the electrons at the gyrofrequency ω_H, also because of the influence of spatial dispersion of the plasma. As we shall see, the branch covers three frequency ranges: ELF, VLF, and LF waves, although it corresponds to only one resonance region of the cold plasma, the LF branch $\omega_2(\theta)$ (see Fig. 3). In the limit when $\omega \ll \Omega_H$, this electron wave is called a fast magnetoacoustic or modified Alfvén wave:

$$\omega = kV_A[1 + (V_A/c)^2]^{-\frac{1}{2}},$$
$$n_{20} = n_A(1 + \omega/\Omega_H)^{-1}, \tag{1.40}$$
$$\kappa_{20} = \tfrac{1}{2}n_{20}\nu\omega[\omega_H(\omega + \Omega_H)]^{-1}.$$

In the frequency region $\omega > \omega_{ij}$ where ω_{ij} are the so-called c r o s s o v e r   f r e q u e n c i e s, which arise because the plasma is multicomponent, i.e., contains several species of ions (see below), this wave is also called an electron whistler. However, the term whistler is usually applied to waves of the low-frequency range $\omega_L < \omega \leqslant \omega_H$, since one uses the corresponding formulas without allowance for the influence of the ions, namely,

$$n_{20}^2 = \omega_0^2[\omega(\omega_H\cos\theta - \omega)]^{-1},$$
$$\kappa_{20} = \tfrac{1}{2}n_{20}\nu(\omega_H\cos\theta - \omega)^{-1}. \tag{1.41}$$

In (1.41), κ_{20} is written down for the region of the near-Earth plasma in which collisional damping of the wave plays a role. This corresponds to $z \lesssim 2000$-3000 km. Higher than this, collisionless damping (see (3.15)) plays the main role. The general expression for n_2^2 of the electron wave 2, like wave 1, is described by (1.37).

The remaining three branches 3, 4, and 5 (two extraordinary n_2 and one ordinary n_1) are high-frequency (HF) waves. For them $n^2 > 0$ only when $\omega > \omega_-$. Their characteristic frequencies are shown in Figs. 1 and 2, namely, the frequencies ω_-, ω_0, ω_+ (see (1.27b)), for which n_{20-}^2, n_{10}^2, and n_{20+}^2 are equal to zero. The frequency range of these waves, when they can propagate in the plasma ($n^2 > 0$), varies in the range $\omega_- \lesssim \omega \to \infty$ and includes the narrow region of high-frequency resonances $\omega_1(\theta) = \omega_0 - (\omega_H^2 + \omega_0^2)^{1/2}$ of a cold plasma (see Fig.3 and equations (1.31) and (1.32)).

To conclude this section, let us consider how the fact that a plasma is a multicomponent one affects the behavior of waves of different modes. This influence is important primarily in the ionospheric regions $z \lesssim 1000$ km, where the principal components are protons, H^+, ions of oxygen, O^+, nitrogen, N^+, and helium, He^+. The presence of several species of ions means, first, that the number of branches 1 of ion ordinary waves is increased in accordance with the number of ions. Accordingly, one observes the same number of ion-cyclotron resonances (see Fig. 4a). The number of values of ω_L also increases accordingly (see Fig. 4b). However, the hybrid frequency corresponding to the resonance branch $\omega_2(\theta)$ (see Fig. 3) is determined in terms of the effective mass of all the ions $(M_{eff})^{-1} = \Sigma \alpha_j / M_j$, where α_j are the relative concentrations of the ions of the species j and $\Sigma \alpha_j = 1$. In this case, it is convenient to write the lower hybrid frequency in the form

$$m/\omega_L^2 M_{eff} = 1/\omega_H^2 + 1/\omega_0^2. \qquad (1.42)$$

Second, because the plasma is multicomponent, the number of zeros of n_1^2 is increased. This complicates the propagation of VLF waves in the ionosphere (see Chapter III). Third, the polarization senses of both waves 1 and 2 depend in a more complicated manner on the frequency; namely, at the intersections of the branches 1 and 2, at the frequencies ω_{12}, ω_{23}, ... , which are called c r o s s o v e r   f r e q u e n c i e s, the polarization senses of the ordinary and extraordinary waves are reversed. In Fig. 4a, left-handed rotation of the electric vector of the wave is denoted by a continuous curve and right-handed rotation by a dashed curve. As we have already mentioned above, these circumstances lead to additional effects in the behavior of VLF waves in the ionosphere (see Chapter III).

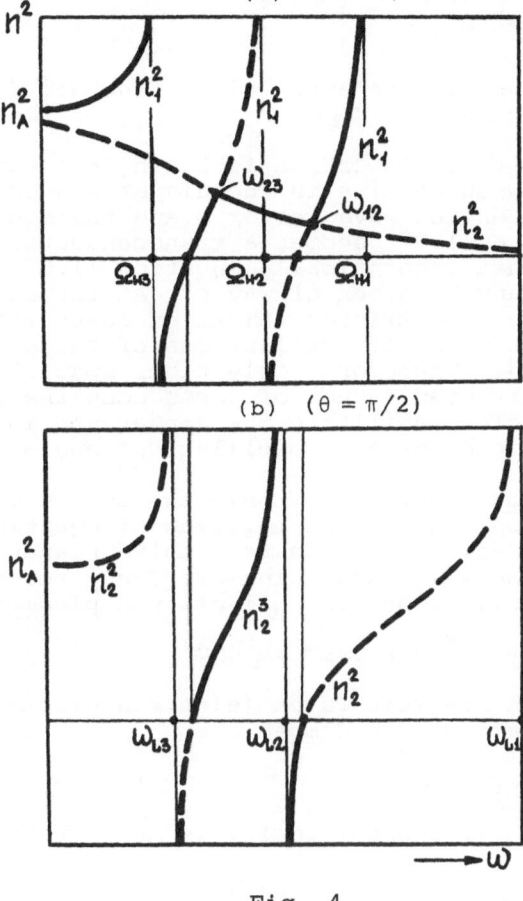

Fig. 4

In a multicomponent plasma, the refractive indices have the form

$$n_1^2 = -\omega_0^2/\omega\omega_H + \Omega_{01}^2/A_1 + \Omega_{02}^2/A_2 + \dots \ ,$$

$$n_2^2 = \omega_0^2/\omega\omega_H - \Omega_{01}^2/B_1 - \Omega_{02}^2/B_2 - \dots \ , \qquad (1.43)$$

$$A_i = \omega(\Omega_{Hi} - \omega), \ B_i = \omega(\Omega_{Hi} + \omega), \ i = 1, \ 2, \dots$$

where the subscripts 1, 2,... denote ions of different species.

§5. Some Forms of Resonances with Allowance for Spatial Dispersion ($T \neq 0$, $T_e \gg T_i$)

When spatial dispersion is taken into account, because all the quantities in the dispersion equation (1.5) depend on the wave vector k and therefore on n as well, equation (1.5) becomes a transcendental equation. This means that theoretically equation (1.5) can have infinitely many branches of waves. At the same time, it is only in a restricted number of cases that the corresponding branches of oscillations of the plasma are weakly damped. Therefore, only under special conditions and in a restricted number of cases does thermal motion of the particles facilitate the appearance in the plasma of weakly damped resonant oscillations and waves.

1. Langmuir Waves. If only the motion of the electrons is considered, the solution of equation (1.5) for $H_0 = 0$ immediately leads to a third branch n_3, which determines the well-known high-frequency resonant longitudinal oscillations in an isotropic plasma:

$$\omega^2 = \omega_0^2 (1 + 3k^2 D^2) = \omega_0^2 + 3\beta_e^2 k^2 c^2 \tag{1.44}$$

the so-called Langmuir waves (electron plasma waves). These waves are weakly damped when

$$\omega/k = v_{ph} \gg v_e, \tag{1.45}$$

i.e., when their phase velocity is much greater than v_e or also if

$$(kD)^2 = (2\pi D)^2/\Lambda^2 \ll 1, \tag{1.46}$$

i.e., the wavelength is much greater than the Debye radius. In (1.44),

$$\beta_e = v_e/c,$$
$$k = \omega n_3/c, \tag{1.47}$$
$$D = (\kappa T/4\pi N e^2)^{\frac{1}{2}} = v_e/\sqrt{2\omega_0}.$$

In the study of high-frequency resonances of a plasma, Landau was the first to discover the collisionless damping mechanism, and it is therefore known as Landau

damping.[6] Collisionless Landau damping is explained
physically by the Cherenkov absorption of longitudinal
waves by electrons. And, since in the considered case
the phase velocity satisfies $v_{ph} \gg v_e$, the only electrons
that play a role in this process are those belonging to
the "tail" of their velocity distribution. Therefore,
the waves are weakly damped. The damping is determined
by the temporal decay rate ($E \sim \exp(-\gamma_e t)$)

$$\gamma_e = \sqrt{\pi/8}\,e^{-3/2}\,(\omega_0/k^3 D^3)\,\exp(-1/2k^2 D^2),\qquad(1.48)$$

and the spatial damping coefficient

$$\kappa_e = \sqrt{\pi/8}\,e^{-3/2}c[3\omega_0 D(kD)^4]^{-1}\exp(-1/2k^2 D^2).\qquad(1.49)$$

Condition (1.46) also shows that the frequency of these
waves is very near ω_0. Their refractive index and phase
velocity are, respectively,

$$n_3^2 = 2(\omega^2 - \omega_0^2)/3\omega^2\beta_e^2,$$

$$v_{ph} = \sqrt{3/2}\,v_e[\omega^2/(\omega^2 - \omega_0^2)]^{\frac{1}{2}} \gg v_e.\qquad(1.50)$$

However, despite the fact that $(\omega^2 - \omega_0^2)/\omega^2 \ll 1$ in the
near-Earth and interplanetary plasma, n_3^2 can take very
large values becuase of the smallness of β_e^2, which varies
approximately in the range 10^{-7}-10^{-5} at different distances from the surface of the Earth.
    2. Electron-Acoustic Waves. It is evident that
weakly damped Langmuir waves occupy a very narrow sec-
tion of frequencies $\Delta\omega = \omega - \omega_0 \ll \omega_0$. This is because the
electron velocity v_e must be small compared with the
phase velocity v_{ph}. A broader spectrum of longitudinal
high-frequency, more strongly damped plasma waves due to
the electron motion and, in a certain range of frequencies, the ion motion as well can be excited in a plasma
if

$$\omega \gg kv_i,\quad v_{ph} \gg v_i,\qquad(1.51)$$

i.e., if the ion thermal velocity v_i is small compared
with the phase velocity v_{ph} of the waves.

    The dispersion equation of these so-called electron-
acoustic waves in an isotropic plasma can be written for
a Maxwellian distribution function in the form[2]

$$\omega^2 = \Omega_0^2 + 2\omega_0^2\alpha_e^2[2\alpha_e I(\alpha_e) - 1],\qquad(1.52)$$

where

$$\alpha_e = \omega/kv_e = v_{ph}/v_e. \tag{1.53}$$

If $\alpha_e \gg 1$, then

$$I(\alpha_e) = \exp(-\alpha_e^2) \int_0^{\alpha} \exp t^2 dt \tag{1.54}$$
$$= (2\alpha_e)^{-1} + (2^2\alpha_e^3)^{-1} + 1 \cdot 3(2^3\alpha_e^5)^{-1} + 1 \cdot 3 \cdot 5(2^4\alpha_e^7)^{-1}$$

In this case, the electron-acoustic waves are Langmuir waves. If $\alpha_e \ll 1$, then

$$I(\alpha_e) \approx \alpha_e - 2\alpha_e^3(3 \cdot 1)^{-1} + 2^2\alpha_e^5(5 \cdot 3 \cdot 1)^{-1} - 2^3\alpha_e^7(7 \cdot 5 \cdot 3 \cdot 1)^{-1}$$
$$+ \ldots \tag{1.55}$$

The spectrum of these waves in an isotropic plasma has two branches that touch at the frequency $\omega > \omega_0$, namely, for

$$\omega_M^2 = \Omega_0^2 + 1.29\omega_0^2. \tag{1.56}$$

At this point ·

$$v_{ph} \approx 1.5v_e. \tag{1.57}$$

At frequencies $\omega < \omega_M$, namely in the frequency range

$$\Omega_0 \leqslant \omega \leqslant \omega_M, \tag{1.58}$$

there lies the branch of slow electron acoustic waves. It is cut off for

$$v_{ph} = 0.924v_e \tag{1.59}$$

at the frequency Ω_0. Thus, this branch covers two frequency ranges: those of the low- and high-frequency waves. The branch of fast electron-acoustic waves lies in the range

$$\omega_M^2 < \omega^2 \leqslant \Omega_0^2 + 1.647\omega_0^2, \tag{1.60}$$

and its phase velocities vary from $v_{ph} = 1.502v_e$ (see (1.56)) to $v_{ph} \gg v_e$. The properties of these waves have not hitherto been analyzed theoretically in detail, especially in a magnetoactive plasma.

3. Ion-Acoustic Waves. Let us now consider weakly damped waves in a nonisothermal plasma, $T_e \gg T_i$, whose phase velocities lie in the intermediate region between the thermal velocities of the ions and the electrons. The spectra of these waves have been well studied theoretically in both an isotropic plasma and a magnetized plasma, when

$$v_i \ll v_{ph}/\cos\theta \ll v_e. \qquad (1.61)$$

In the isotropic case (when $H_0 = 0$, $\cos\theta = 1$) the dispersion equation of these waves, which are frequently called electrostatic waves or Langmuir-Tonks waves, has the form

$$\omega = \omega_{10}^2 = k^2 v_s^2/(1 + k^2 D^2) = \Omega_0^2/\{1 + (k^2 D^2)^{-1}\}, \qquad (1.62)$$

where the subscript 10 is appended to distinguish them from the corresponding two branches of waves in a magnetoactive plasma. In the limiting case, namely, when

$$k^2 D^2 = D^2/\Lambda^2 \ll 1, \qquad (1.63)$$

i.e., when the waves are sufficiently long

$$\omega = k v_s, \qquad (1.64)$$

where $v_s = \sqrt{\kappa T_e/M}$ is the velocity of nonisothermal sound. Equation (1.64) is like the equation of sound waves. Therefore, waves described by equations (1.62) and (1.64) are called ion-acoustic waves. It is also important to point out that when $(kD)^2 \gg 1$

$$\omega_{10} \to \Omega_0, \qquad (1.65)$$

i.e., ω_{10} does not depend on the wave vector k, and low frequency ion Langmuir oscillations arise in the plasma.

To describe these wave in a magnetized plasma, one can use the approximate dispersion equation

$$
\begin{aligned}
1/\omega^2 + \tan^2\theta/(\omega^2 + \Omega_H^2) &= (\cos^2\theta)^{-1} (1/k^2 v_s^2 + 1/\Omega_0^2) \\
&= (\cos^2\theta)^{-1} [(1 + k^2 D^2)/k^2 v_s^2] \\
&= (\cos^2\theta)^{-1} [1 + (k^2 D^2)^{-1}]/\Omega_0^2,
\end{aligned}
\qquad (1.66)
$$

which determines the two branches $\omega_1(k,\theta)$ and $\omega_2(k,\theta)$, fast and slow ion-acoustic waves:

$$\omega_{12}^2 (k\theta) = \tfrac{1}{2}[(\omega_{i0}^2 + \Omega_H^2) \pm \{ (\omega_{i0}^2 + \Omega_H^2)^2 - 4\omega_{i0}^2 \Omega_H^2 \cos^2\theta \}^{\tfrac{1}{2}}].$$

$$(1.67)$$

It follows directly from (1.67) that $\omega_1(k,\theta) = \omega_{10}$ for $H_0 = 0$ ($\cos\theta = 1$). It is also obvious from condition (1.61) that formulas (1.66) and (1.67) are to be used with care as θ approaches $\pi/2$ and as $\omega \to \Omega_H$. Corresponding criteria for the applicability of these formulas are given in ref. 3 and have the form

$$\theta^2 \gg 2kv_i |\Omega_H^2 - \omega_{i0}^2|/\Omega_H \omega_{i0}^2 ,$$
$$|\Omega_H - \omega|/kv_i \cos\theta \gg 1,$$

$$(1.68)$$

With these reservations, the spectra of the fast and slow ion-acoustic waves, namely the depdences $\omega_{12}(\theta)$ and $\omega_{12}(k)$, are shown schematically in Fig. 5. For simplicity, it is assumed in Fig. 5 that $kv_s > \Omega_H$, although this case is seldom realized. It can be seen that the ranges of the fast and slow ion-acoustic waves can vary in the different limiting cases ($kD \gg 1$ or $\omega_{10} \gtrless \Omega$) in the following intervals:

$$\Omega_H < \omega_1 < (\Omega_0^2 + \Omega_H^2)^{\tfrac{1}{2}},$$
$$0 < \omega_2 < \Omega_H.$$

$$(1.69)$$

Their refractive indices are

$$n_{12}^2 = (c^2\Omega_0^2/v_s^2) \{\Omega_0^2 (\omega_{12}^2 - \Omega_H^2 \cos^2\theta)(\omega_{12}^2 - \Omega_H^2)^{-1} - \omega_{12}^2\}^{-1},$$
$$n_{10}^2 = (c\Omega_0^2/v_s^2)(\Omega_0^2 - \omega^2)^{-1},$$

$$(1.70)$$

and the decay rate and damping coefficient of these waves, due largely due to absorption by electrons, are

$$(\gamma_e)_{12} = \sqrt{\pi m/8M}\,\omega_{12}^4 (kv_s)^{-3}|\cos\theta|^{-1}$$
$$\times [\cos^2\theta + \omega_{12}^4 \sin^2\theta (\omega_{12}^2 - \Omega_H^2)^{-1},$$

$$(1.71)$$

$$\kappa_{e1} = \tfrac{1}{2}\sqrt{\pi}c(v_e |\cos\theta|)^{-1} (\omega_1^2 - \Omega_H^2)[(\omega_1^2 - \Omega_H^2) + (\omega_1^2 - \omega_{10}^2)]$$
$$\times \{ (\omega_1^2 - \Omega_H^2 \cos^2\theta)[\cos^2\theta (\omega_1^2 - \Omega_H^2) + \omega_1^4 \sin^2\theta]^{-1},$$

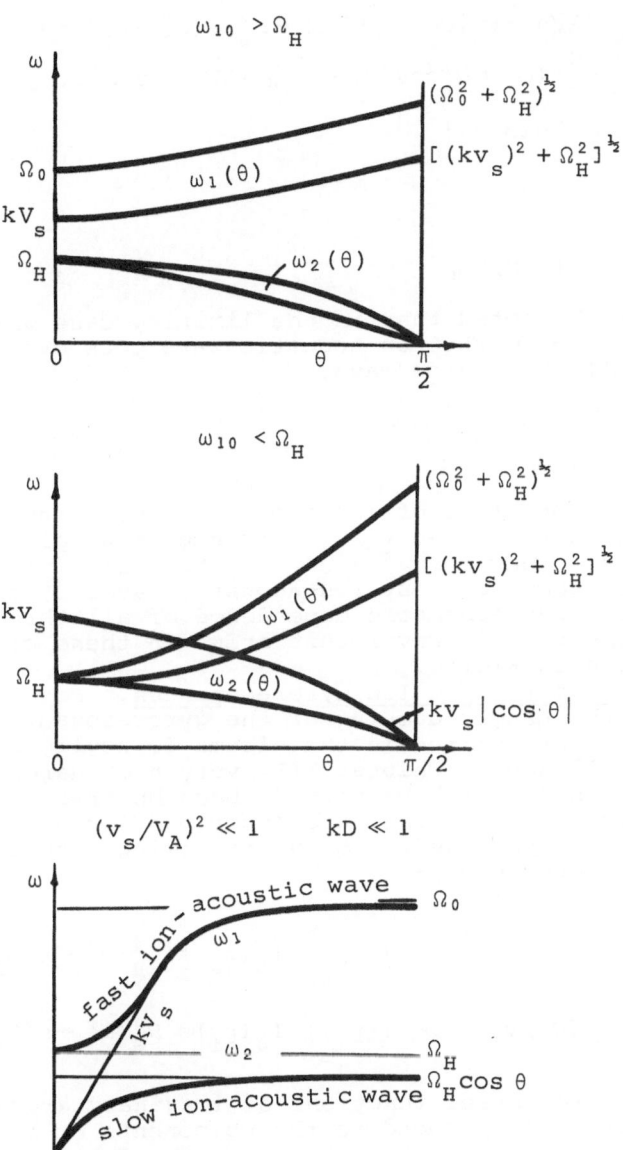

Fig. 5.

$$\kappa_{e2} = \tfrac{1}{2}\sqrt{\pi}c\,(v_e|\cos\theta|)^{-1}(\omega_2^2 - \Omega_H^2)\,[\,(\omega_2^2 - \Omega_H^2) + (\omega_2^2 - \omega_{10}^2)\,]$$

$$\times\,\{\,[\,(\omega_2^2 + \Omega_H^2\cos^2\theta) - (\omega_{10}^2 + \Omega_H^2)\,][\cos^2\theta\,(\omega_2^2 - \Omega_H^2)$$

$$+\,\omega_2^4\sin^2\theta\,]\,\}^{-1}.$$

When $\omega_{12} \sim kv_s \sim \Omega_H$,

$$(\gamma)_{12} \sim (m/M)^{\frac{1}{2}}\Omega_H. \tag{1.72}$$

It should be noted that in the limiting case when $kv_s \ll \Omega_H$ and $kD \ll 1$ the ion-acoustic wave goes over into a slow ELF transverse wave:

$$\omega_2 = kv_s\cos\theta,$$

$$\gamma_{e2} = (\pi m/8M)^{\frac{1}{2}}\omega_2.$$

The properties of ion-acoustic waves change appreciably with the ratio v_s/V_A. In Fig. 6 we give the dependences $\omega_{12}(k)$ for $v_s^2/V_A^2 \ll 1$. When $v_s/V_A \sim 1$ or $v_s/V_A \gg 1$, the branch of fast ion-acoustic waves ω_1 is split into two. The schematic dependence of all three branches of ion-acoustic waves that arise in these cases is also shown in Fig. 6.

4. Electron and Ion Gyroresonances. For angles $\theta \approx \pi/2$, the Landau damping of the gyroresonant longitudinal waves excited in the plasma is negligibly small. As we shall see in Chapter III, very high harmonics of the gyroresonances have already been observed in the near-Earth plasma. They are described for electrons and ions, respectively, by the following dispersion equations (Bernstein[7]):

$$k_e^2 = 4\,(\omega_0^2/v_{e\perp}^2)\,\exp\,(-p_e)\sum_{s=1}^{\infty} I_s\,(p_e)\,s^2\omega_H^2\,(\omega^2 - s^2\omega_H^2)^{-1},$$

$$k_i^2 = 4\,(\Omega_0^2/v_{i\perp}^2)\,\exp\,(-p_i)\sum_{s=1}^{\infty} I_s\,(p_i)\,s^2\Omega_H^2\,(\omega^2 - s^2\Omega_H^2)^{-1} \tag{1.73}$$

where I_s are Bessel functions of imaginary argument, the subscript $\perp$ is appended to the components of the corresponding quantities that are perpendicular to the direction of H_0, and

$$p_e = k_\perp^2 v_{e\perp}^2/2\omega_H^2, \tag{1.74}$$

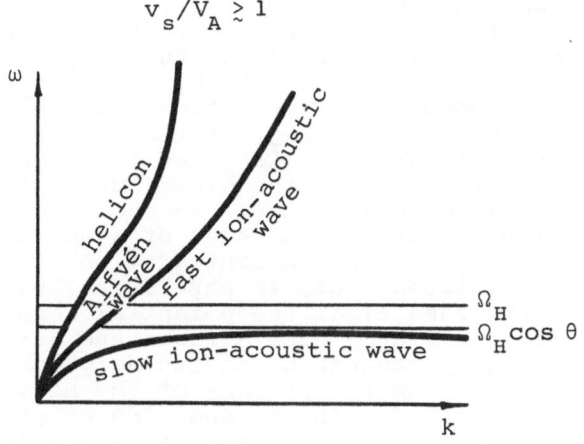

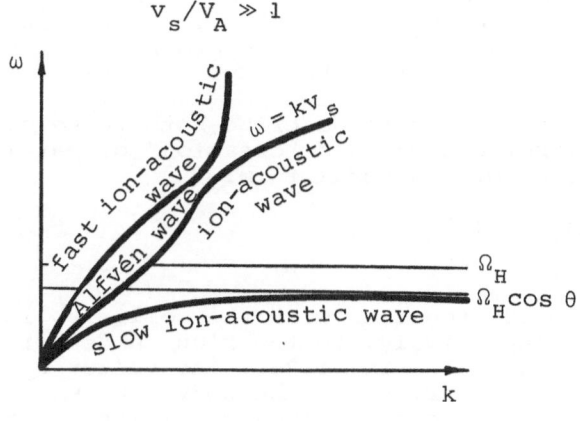

Fig. 6

$$p_i = k_\perp^2 v_{i\perp}^2 / 2\Omega_H^2.$$ (1.74)

§6. Some Remarks on the Conditions on the Surfaces of Bodies Moving in the Plasma

The phenomena that arise in a plasma in the neighborhood of moving bodies naturally depend on the form

of the body and the physical properties of its surface.
With regard to the form of the body, it is clear that
this plays a role only fairly near the body; this influence will be illustrated in §9 by some examples.
However, the structure of the surface and the matter
from which it is made play a greater role. First, depending on the nature of the "reflection" of particles
by the surface of the body, the distribution function of
the plasma particles may change. Second, there may be
"creation" of new particles because of evaporation of
the surface of the body and disintegration of the surface under the influence of, in particular, particle
streams. At the same time, the potential acquired by a
body moving in the plasma also affects the structure and
the nature of the perturbation in its neighborhood. The
boundary conditions on the surface of the body may also,
for example, be decisive in the question of stability
and the excitation of oscillations of the plasma. On
the other hand, they have little influence on the scattering of radio waves on the wake of the body, in which
the entire region of the perturbation participates as a
whole, and which is primarily determined by regions of
the wake fairly far from the body.

It is convenient to describe both these groups of
phenomena ("reflection" and "creation" of particles) by
means of terms of a general form:

$$A_e(r_s, v_1, v)\, \delta(s),$$
$$A_i(r_s, v_1, v)\, \delta(s), \tag{1.75}$$

which are added, together with the collision integrals
Y_e and Y_i, respectively, to the right sides of equations
(1.11) (see ref. 5). In (1.75), the vector r_s determines
the point on the surface of the body, and v_1 and v are,
respectively, the velocities of particles before and
after collision with the body. Depending on the physical
formulation of the problem, the functions A_e and A_i
have the same or different form and can, in particular,
include several terms that describe each particular type
of process. The delta function $\delta(s)$ in (1.75) describes
the fact that the functions A_e and A_i are nonvanishing
only on the surface of the body. The physical meaning
of these functions is clear. It is of course obvious
that $A\delta(s)$ has the same dimensions as $\partial f/\partial t$, and therefore

$$\int A\, d\vec{v}^3 = J \tag{1.76}$$

determines the change in the flux of particles in unit time (J cm^{-2}·sec^{-1}) due to the influence of the surface of the body.

The above three groups of questions -- reflection and creation of particles and potential of the body -- which determine the influence of the surface of the body on the way in which the plasma flows around the body, have as yet been little studied in many respects. This is because it is theoretically very difficult and sometimes impossible to perform the corresponding calculations; moreover, one frequently does not even know how the problem should be posed, since one is frequently in ignorance of a number of basic data, which can only be obtained theoretically. At the present time it is therefore impossible to discuss these questions fully, and we shall therefore restrict ourselves to some brief comments.

1. Reflection of Particles. The term reflection covers phenomena that have quite different physical bases. S p e c u l a r   r e f l e c t i o n of particles, for which the angle of incidence is equal to the angle of reflection and $|v_1| = |v|$ is realized only from an absolutely smooth dielectric surface. In the case of a spherical surface we have, for example, for ions the boundary conditions[5]

$$A_i \delta(s) = r \cdot V r^{-1} f_i (r, v - 2r(r,v)r^{-2}) \delta(r - \rho_0) \qquad (1.77)$$

for $r \cdot v > 0$ and

$$A_i \delta(s) = r \cdot V r^{-1} f_i (r, v) \delta(r - \rho_0) \qquad (1.77a)$$

for $r \cdot v < 0$, where $V_0 \gg v_i$ is the velocity of a sphere of radius ρ_0. However, one can also have cases of reflection of e l a s t i c - d i f f u s e type, when the directions of the reflected particles are equally probable, while the moduli of their velocities are conserved. In this case one generally speaks of s c a t t e r i n g of particles, as in the case of i n e l a s t i c reflection (when one speaks of p a r t i a l   a c c o m o d a t i o n), when the directions are equally probable and the velocity moduli are decreased because particle give energy to the surface. In the case of complete accomodation of the particles, i.e., when they are completely a b s o r b e d by the surface,

$$A_i \delta(s) = 0,$$

$$A_i \delta(s) = r \cdot V_0 r^{-1} f_i \delta(r - \rho_0)$$

(1.78)

respectively, if $r \cdot v > 0$ or $r \cdot v < 0$.

2. "Creation" of Particles. The plasma surrounding
the body must be continuously filled with the particles
of which the surface of the body consists as a result
of e v a p o r a t i o n and e r o s i o n due to the bombardment
of the body by particle streams or meteor matter. Particles are also "created" as a result of electronic and
ionic photoemission and other processes. Neutral atoms
of molecules that leave the surface are ionized very
slowly. According to different estimates, the i o n i-
z a t i o n   t i m e in the media in which we are interested
is

$$\tau_i \approx 10^7 \text{ sec,}$$

(1.79)

and the velocity with which the particles leave the body
is

$$v_j = \sqrt{2\kappa T_{sj}/M_j} \approx 10^4 \text{ cm/sec}$$

(1.80)

(the index j distinguishes the particle). Thus, the
evaporated particles first move away slowly from the
body, and it is only at large distances that they acquire the thermal velocity of the surrounding medium,
diffusing rapidly. The time required for a particle to
move away from the surface of the body by a distance of,
for example, $r_j \approx 10^2$ cm is

$$\tau_j \approx r_j/v_j \sim 10^{-2} \text{ sec.}$$

(1.81)

Therefore, in this region, the ratio of the density N_j
of charged particles to the density n_j of neutral particles is negligibly small:

$$N_j/n_j \sim \tau_j/\tau_i \sim 10^{-9}.$$

(1.82)

At the same time, the fluxes of created particles measured on satellites show that around the body itself the
number of "created" particles n_j may be fairly high.
Therefore, in regions of the plasma where $n_j \gg n_0$ i.e.,
the density of neutral particles of the natural plasma
is negligibly small, the created particles may play an
important role in the processes taking place around the
body. This circumstance must evidently be borne in mind

in a number of cases when one is considering different experimental data. Thus, in one series of experiments (McKeown[8]) it was shown that at altitudes $z \sim 216$–810 km the flux of particles lost by a gold plate placed at right angles to the incident stream (to the vector V_0) varied in the range

$$J_j = (\overline{nv})_j \approx 10^7\text{–}10^{10} \text{ atoms.cm}^{-2}.\text{sec}^{-1}, \qquad (1.83)$$

which corresponded under the experimental conditions to a rate of evaporation $\leq 5 \cdot 10^{-6}$ of gold atoms per particle of the oncoming stream. For other metals (aluminum, zinc, iron, magnesium, lithium) the value observed for $T_j \sim 10^2$ –10^3 was

$$J_j \approx (\overline{nv})_j \approx 10^{10}\text{–}10^{14} \text{ .atoms.cm}^{-2}.\text{sec}^{-1}. \qquad (1.84)$$

The rate of sublimation in vacuum of polymers, nylon, sulfides, and vinyl chloride reaches $3 \cdot 10^{-9}$ of the weight of matter per second. Thus, if we use the data (1.83) and (1.84), we find that for $v_j \approx 10^4$ cm/sec we can expect in different cases

$$n_j \approx (\overline{nv})_j / v_j \sim 10^3\text{–}10^{10} \text{ cm}^{-3}. \qquad (1.85)$$

3. Potential of the Body. The question of the potential ϕ_0 acquired by a body as it moves in the near-Earth is very important. In a number of experiments, knowledge of ϕ_0 is actually decisive and determines the accuracy with which one can interpret the results of measurements. On the other hand, as we have already pointed out, the potential of the body affects the manner in which the plasma flows around it, expecially in the near zone of the perturbation of the plasma. It is hardly possible to calculate theoretically the potential ϕ_0 accurately because of the complexity of the geometrical and electrical structure of the surface of the body and the absence of some of the basic data on the interaction between the matter of which the body consists and the streams of particles and radiation incident on it. Let us consider briefly simple estimates for ϕ_0.

At any point on the surface of a weakly conducting body the potential is determined by the condition of equality of the number of electrons and ions absorbed by the surface in unit time. And, since $v_e \gg v_i$, the body must acquire a negative charge; for suppose that electrons and singly charged ions are incident on the body and that $N_e \approx N_i$ and $T_e \approx T_i$. At an arbitrary point of the plasma, the ratio of the fluxes is

$$J_e/J_i \sim v_e/v_i \gg 1. \tag{1.86}$$

Therefore, when the surface of the body is bombarded by these particles it will be charged as long as the electron and ion fluxes do not become equal at a given point s. But that can only happen if the flux of electrons to the body is decreased, i.e., if the body is negatively charged and repulses the electrons.

Let us now consider first the case of a body at rest. Then the density of the electron flux at a point s on its surface can be written in the form

$$J_{es} = J_{e0} \exp \{-|e\phi(s)|/\kappa T\}, \tag{1.87}$$

where $J_{e0} = Nv_e/2\sqrt{\pi}$ cooresponds to the electron flux unperturbed by the body, i.e., when $\phi_0 = 0$. The ion flux depends in a more complicated manner on the potential. However, in the limiting case for the maximally expected values of ϕ_s we can assume that

$$J_{is} \approx J_{i0} \approx N_0 v_i/2\sqrt{\pi} \tag{1.88}$$

i.e., that J_{is} is equal to the unperturbed flux J_{i0} of the ions. Assuming now that the coefficients of reflection of the ions and the electrons from the body are, respectively, ρ_i and ρ_e, we can determine ϕ_s from the equation

$$J_{is}(1 - \rho_i) = J_{es}(1 - \rho_e), \tag{1.89}$$

or, using (1.87) and (1.88), we obtain

$$|\phi_s| = (\kappa T/e) \ln \{v_e(1 - \rho_e)/v_i(1 - \rho_i)\}. \tag{1.90}$$

Hence, for a completely absorbing body in a state of quasirest ($v_i \gg V_0$), when ρ_i, $\rho_e \ll 1$, it follows that in the media in which we are interested (see Tables 1 and 2 in §2) that

$$|\phi_s| \sim 1\text{-}2 \text{ V.} \tag{1.91}$$

For a rapidly moving body ($V_0 \gg v_i$) one must, when determining the potential of the front surface of the body, substitute into (1.89) the ion flux (1.88) in the form

$$J_i \sim NV_0 \cos \zeta_0, \tag{1.92}$$

where ζ_0 is the angle of incidence of the particle on the body. When $\cos \zeta_0 \sim 1$ we therefore have

$$|\phi_s| = (\kappa T/e)\ \ln\left\{v_e(1 - \rho_e)/V_0(1 - \rho_i)\right\} \sim 0.5\text{-}1\ \text{V}. \tag{1.93}$$

It is difficult to calculate ϕ_s behind the body, since there are no sufficiently simple or exact formulas for the particle fluxes in this region. The potential of a metallic body must be constant, so that the potential $\phi_0 \approx \phi_s$ on its front surface is established all over the surface. However, for a dielectric body with inhomogeneous surface, the potential must vary strongly from point to point due to the changes, in particular, in the coefficients of reflection, the influence of different emission processes, and so forth. In this case, the potential on the rear surface will be increased because the ion flux is decreased compared with the electron flux. These circumstances may also explain the fact that in different experiments one has observed values of ϕ_0 that appreciably exceed the values estimated by means of (1.91) and (1.93) (see, for example, Sharp, Hanson, McKiblin[9]). These experiments were carried out on satellites that usually had fairly complicated geometrical and electrical structures, their surfaces having fairly "sharp" protuberances, conducting regions alternating with dielectric regions, etc. In such cases, charge may accumulate at various parts of the surface of the body, this resulting in an appreciable increase in the potential ϕ_0. This question has not however been hitherto elucidated and, as we have mentioned above, is of great importance, in particular, for the correct interpretation of the results of different measurements of probe type. We mention here that recently some very interesting experiments have been made with the Gemini-Agena system, in which investigations were made of the properties of the wake of the body in the ionosphere (Troy, Medved, Samir[10]). In these experiments, values of the potential of the order $\phi_0 \sim -0.5$ V were obtained, in good agreement with the estimate (1.93) made above. Similar good agreement with (1.93) was obtained earlier in the measurements of ϕ_0 on Explorer 31 (Samir, Wrenn;[35] see also ref. 165).

§7. Group Velocity

It is important to know the group velocity **u**, which determines both the velocity of propagation of the envelope of the wave packet -- the signal -- or the ve-

locity of transport of the signal energy, u_ε, both for
the analysis of different cases of propagation or excitation of waves and for the interpretation of the results of experiments and also when one is analyzing a
number of processes in the neighborhood of bodies moving in the plasma. We shall consider here the basic
properties of the vector of the group velocity in the
whole of the relevant range of frequencies.

In the general case, the velocity of transport of
energy in a dispersive medium is

$$u_\varepsilon = s_H/w, \tag{1.94}$$

where s_H is the energy flux density averaged over a period and w is the mean bulk energy density of the wave
packet at the carrier frequency. In dispersive nonab-
sorbing anisotropic and isotropic media, in particular,
in the plasma regions in which we are interested, the
velocity of energy transport is equal to the expression
that is usually used for the group velocity, namely

$$u_\varepsilon = u = \partial\omega/\partial k. \tag{1.95}$$

In absorbing media, the relation (1.95) is not satisfied,
and to calculate u_ε in each individual case it is necessary to calculate w. This is frequently a complicated
problem.

The modulus of the vector of the group velocity
(Al'pert,[11,12] Stix[2]) is

$$|u| = |\partial\omega/\partial k| = \alpha(\partial\omega/\partial k_x) + \beta(\partial\omega/\partial k_y) + \gamma(\partial\omega/\partial k_z), \tag{1.96}$$

where $\alpha = \sin\theta$, β, and $\gamma = (k\cdot H_0) = \cos\theta$ are the angular
components of the wave vector k. In a collisionless
plasma, u lies in the plane (k, H_0), and to simplify the
calculations it is therefore assumed that $\beta = 0$. It then
follows that

$$|u| = c[(n + \omega\partial n/\partial\omega)\cos\psi]^{-1},$$

$$\tan\psi = n^{-1}\sin\theta\partial n/\partial(\cos\theta),$$

$$\tan\alpha = \tan(\theta - \psi) = (\partial\omega/\partial k_x):(\partial\omega/\partial k_z) \tag{1.97}$$

$$= \frac{\tan\theta\partial(n\cos\theta)/\partial(\cos\theta)}{\partial(n\cos\theta)/\partial(\cos\theta) - (\cos\theta)^{-1}\partial n/\partial(\cos\theta)}$$

where ψ and α are the angles between the vector $\mathbf{u}$ and, respectively, the vectors $\mathbf{k}$ and $\mathbf{H}_0$, i.e., $\cos \psi = (\mathbf{k} \cdot \mathbf{u})$, $\cos \alpha = (\mathbf{u} \cdot \mathbf{H}_0)$. Note that

$$\cos \psi = [1 + n^{-2} \sin^2 \theta (\partial n / \partial \cos \theta)^2]^{-\frac{1}{2}},$$

$$n + \omega \partial n / \partial \omega = \partial (\omega n) / \partial \omega = n_g. \tag{1.98}$$

The quantity n_g is usually called the group refractive index.

1. High-Frequency Waves. In the range of high-frequency transverse electromagnetic waves $\omega_H \leqslant \omega \to \infty$ (branches 3, 4, and 5 in Fig. 1) one readily obtains from (1.25) for the determination of the modulus of the group velocity and the angles ψ and α the following values of the partial derivatives of the refractive indices n_{12}, which occur in (1.97):

$$\partial n_{12} / \partial (\cos \theta) = n_{12} (1 - n_{12}^2) \cos \theta (\omega_H / \omega)^2$$
$$\times \{ 2 (1 - \omega_0^2 / \omega^2) (\omega_0^2 / \omega^2 - 1 - n_{12}^2) + (1 - n_{12}^2) (1 - \cos^2 \theta) n_{12}^2 \}^{-1},$$

$$\partial n_{12} / \partial \omega = (1 - n_{12})^2 [n_{12} \omega (1 - \omega_0^2 / \omega^2)]^{-1}$$
$$\times \{ 1 - \tfrac{1}{2} (1 - n_{12}^2) (\omega_0^2 / \omega^2)^{-1} ((2 (2 - \omega_0^2 / \omega^2) - (\omega_n^2 / \omega^2) \sin^2 \theta)$$
$$\times [(\omega_H^4 / \omega^4) \sin^4 \theta + 4 (\omega_H^2 / \omega^2) (1 - \omega_0^2 / \omega^2) \cos^2 \theta]^{\frac{1}{2}} \quad (1.99)$$
$$- (\omega_n^2 / \omega^2) [(\omega_H^2 / \omega^2) \sin^4 \theta - 8 (1 - \omega_0^2 / \omega^2) \cos^2 \theta])$$
$$\times [(\omega_H^4 / \omega^4) \sin^4 \theta + 4 (\omega_H^2 / \omega^2) (1 - \omega_0^2 / \omega^2)^2 \cos^2 \theta]^{\frac{1}{2}} \}.$$

The main property of the vector of the group velocity in this range of frequencies is that its direction may be very different from that of the of the wave front $\mathbf{k}$. For different values of ω_0 / ω and ω_H / ω, the angle ψ may vary in wide intervals, namely, from 0 to 90° (see refs. 11 and 15). Thus, high-frequency waves are not always forced onto the direction of the vector $\mathbf{H}_0$ in a magnetoactive plasma.

2. Low-Frequency Waves. For electron whistlers (branch 2 in Figs. 1 and 2) in the frequency range $\omega_L \ll \omega \lesssim \omega_H$ it follows from (1.42) and (1.97) that

$$|\mathbf{u}_2| = c n_g^{-1} [1 + \tfrac{1}{4} \omega_H^2 \sin^2 \theta (\omega_H \cos \theta - \omega)^{-2}]^{\frac{1}{2}}.$$

$$\tan \psi = -\tfrac{1}{2} \tan \theta \cdot \omega_H \cos \theta (\omega_H \cos \theta - \omega)^{-1},$$

$$\text{(1.100)}$$
$$\tan \alpha = \tan \theta \cdot (\omega_H \cos \theta - 2\omega) \cos \theta (\omega_H \cos^2 \theta - 2\omega \cos \theta + \omega_H)^{-1},$$

$$n_g = \tfrac{1}{2} n_2 \omega_H \cos \theta (\omega_H \cos \theta - \omega)^{-1}.$$

Analysis of formulas (1.100) leads to the following important properties of low-frequency waves. The vector of the group velocity always lies between H_0 and k (tan $\psi < 0$). This means that the transport path of a packet of these waves is forced onto the line of force of the magnetic field. The maximal value α_M of the angle by which the vector u can deviate from H_0 corresponds to the value

$$\cos \theta_M = [1 - (\omega/\omega_H)^2]^{\tfrac{1}{2}}/\sqrt{3} + \omega/\omega_H. \qquad \text{(1.101)}$$

When $\omega/\omega_H \ll 1$,

$$\tan^2 \alpha_M = 1/8, \quad \alpha_M = 19.5^\circ. \qquad \text{(1.102)}$$

This value of the angle was already determined for whistlers by Storey.[13] For finite values of ω/ω_H the angle α_M decreases and, for example, when $(\omega/\omega_H)^2 = 0.1$ we have $\alpha_M = 10.2^\circ$.

3. VLF and ELF Waves. On the transition to lower frequencies, when it is already necessary to take into account the influence of the ions, the vector u of the branch 2 of waves is more nearly coincident with k. For example, for $\omega \lesssim \Omega_H$ one can, using (1.37), readily find that

$$|u_2| = c n_A^{-1} [(1 + 3\cos^2 \theta)/(1 + \cos^2 \theta)]^{\tfrac{1}{2}},$$
$$\tan \psi = -\tan \theta \cdot \cos^2 \theta (1 + \cos^2 \theta)^{-1}, \qquad \text{(1.103)}$$

from which there follows the maximal value $\alpha_M = 74.2^\circ$ (see ref. 12). We see that the propagation of these waves has become much more isotropic. In the limit $\omega \ll \Omega_H$, we have $\partial n_2/\partial (\cos \theta) = 0$, i.e., $\psi = 0$. This means that the modified Alfvén ELF wave propagates isotropically in the plasma.

In contrast, the ion ELF wave (branch 1 in Fig. 1) is strongly guided by the magnetic field. In the frequency range $\omega \lesssim \Omega_H$ we find that

$$|\mathbf{u}_\perp| = c\left[\tfrac{1}{2}n_A(2 - \omega/\Omega_H)(1 - \omega/\Omega_H)^{-3/2}\right]^{-1}$$
$$\times \left[2(1 + 2\cos^2\theta + \cos^6\theta)(1 + \cos^2\theta)^{-3}\right]^{\tfrac{1}{2}},$$

$$\tan\psi = -\tan\theta \cdot (\cos^2\theta)^{-1},$$

$$\tan\alpha = \tan\theta \cdot \cos^4\theta(1 + \cos^4\theta)^{-1}.$$

(1.104)

It follows from (1.104) that in this range of frequencies the maximal value of the angle between $\mathbf{u}$ and H_0 is

$$\alpha_M = 12.3^\circ.$$

(1.105)

In the limit $\omega \ll \Omega_H$, the group velocity of the ELF Alfvén wave is collinear with the vector H_0, $\alpha_M \to 0$. For both ELF waves when $\omega \ll \Omega_H$

$$|\mathbf{u}|_{12} = c(1 + \omega/\Omega_H)^{3/2}\left[\tfrac{1}{2}n_A(2 \pm \omega/\Omega_H)\right]^{-1}.$$

(1.106)

4. Langmuir and Magnetoacoustic Waves. For longitudinal electron Langmuir HF waves we obtain from (1.44) the expression[12]

$$u = \partial\omega/\partial k = \sqrt{3\kappa T_e/m}\left[(\omega^2 - \omega_0^2)/\omega^2\right]^{\tfrac{1}{2}}$$
$$= \sqrt{3/2}\left[(\omega^2 - \omega_0^2)/\omega^2\right]^{\tfrac{1}{2}}v_e,$$

(1.107)

from which it can be seen that $u \ll v_e$ since $(\omega^2 - \omega_0^2)/\omega^2 \ll 1$ (see §4).

The group velocity of the ion-acoustic waves (ion Langmuir waves) is

$$u = c^3/n^3v_s^2 = c(\Omega_0^2 - \omega^2)/n\Omega_0^2 = v_{ph}(\Omega_0^2 - \omega^2)/\Omega_0^2.$$ (1.108)

Its value is nearly equal to the phase velocity v_{ph} when $\omega \ll \Omega_0$, i.e., when the waves are sufficiently long, $kv_s \ll \Omega_0$ (see §4). For short waves in the upper part of the frequency range, when $\omega \to \Omega$ and the ion-acoustic waves are transformed gradually into longitudinal ion Langmuir oscillations, the group velocity satisfies $u \ll v_{ph}$. In the limit, of course, $u \to 0$ -- the dependency of the frequency on the wave vector disappears.

For fast and slow ion-acoustic waves in a magnetoactive plasma ($H_0 \neq 0$) we obtain from (1.66) (see ref. 12)

$$|\mathbf{u}|_{12} = (c^3/n_{12}^3v_s^2)\left\{\left[(\omega_{12}^2 - \Omega_H^2)^2 + (v_s^4/c^4)n^4\Omega_H^4\sin^2\theta\cos^2\theta\right]\right.$$
$$\times\left[\omega_{12}^2(\omega_{12}^2 - \Omega_H^2\cos^2\theta) - \Omega_H^2(\omega^2 - \Omega_H^2)\right]^{-1}\right\}^{\tfrac{1}{2}}.$$ (1.109)

$(\tan \alpha)_{12} = \tan \theta [1 - \Omega_H^2 \omega_{12}^{-2} (1 + \Omega_H^2/\Omega_0^2 - \omega_{10}^2/\Omega_0^2)^{-1}]$.

(We assume in this book that $\Omega_H < \Omega_0$ or $\Omega_H \ll \Omega_0$).

It follows from inspection of (1.109) that the vector $\mathbf{u}$ of the fast wave, whose spectrum is $\omega_1 = \Omega_H - (\Omega_0^2 + \Omega_H^2)^{\frac{1}{2}}$, is originally forced onto H_0. However, with increasing frequency, the sign of α changes and in the limit $\omega_1 \to \Omega_0$ the vector $\mathbf{u}$ approaches $\mathbf{k}_0$. The direction of the vector of the group velocity of the slow wave ($\omega_2 = 0 \to \Omega$) nearly coincides with H_0 right up to values of the angle θ that differ little from $\pi/2$.

We also give here formulas for the group velocity for magnetoacoustic transverse ELF waves. They are formed in a nonisothermal plasma by the transformation of a modified fast Alfvén wave. When $T_e \gg T_i$, instead of the two waves (1.36) and (1.41), there arise three branches of waves. One of them is the ordinary slow Alfvén wave,

$$\omega_1 = k_1 V_A \cos \theta . \qquad (1.110)$$

However, because of the influence of the electron motion, two fast waves are formed:

$$\omega_{23} = k V_{23} ,$$

$$V_{23}^2 = \tfrac{1}{2} \{ (V_A + v_s^2) \pm [(V_A^2 + v_s^2) - 4 V_A^2 v_s^2 \cos^2 \theta]^{\frac{1}{2}} \} . \qquad (1.111)$$

Their group velocities are

$$|\mathbf{u}| = c n_{23}^{-1} \{ 1 + 4 V_A^2 v_s^2 \cos^2 \theta \sin^2 \theta (V_{23}^4 B)^{-1} \}^{\frac{1}{2}} ,$$

$$(\tan \alpha)_{23} = \tan \theta \cdot \{ [(V_A^2 + v_s^2) - 4 V_A^2 v_s^2 \cos^2 \theta]^{\frac{1}{2}} + C \} (B^{\frac{1}{2}} - C)^{-1} ,$$

$$B = (V_A^2 + v_s^2)^2 - 4 V_A^2 v_s^2 \cos^2 \theta , \qquad (1.112)$$

$$C = 2 V_A^2 v_s^2 V_{23}^{-2} \cos^2 \theta .$$

It can be seen from (1.112) that for both waves the vector $\mathbf{u}$ is not forced onto the vector H_0, but rather deviates from it: $\alpha > \theta$.

CHAPTER II

PLASMA FLOW AROUND MOVING BODIES

§8. Brief Characterization of the Solved Theoretical Problems and Some Experiments

In the general case, theoretical investigation of effects in the neighborhood of bodies moving in a plasma requires a selfconsistent solution of the system of kinetic equations (1.11) and the Poisson equation (1.12) in conjunction with the boundary condition (1.75). As we have already pointed out, this system of equations simplifies if one takes into account the fact that the electrons have a Maxwell-Boltzmann distribution (see (1.16)). Then the Poisson equation (1.12) can be replaced by (1.17). However the simplified system is still very complicated. This explains why only a few classes of problems have been adequately analyzed; primarily, these are the ones for which the equations can be linearized. Usually, linearization reduces to imposing restrictions on the value of the potential of the perturbed region of the plasma and the potential of the body (they are required to be sufficiently small) or on the distance one is from the body (it is required to be sufficiently large). Some problems have however been solved without corresponding restrictions or when only one of them has been fulfilled. In this case, the equations are already of nonlinear type. An important circumstance is also the imposing of conditions on the linear size ρ_0 of the body, its commensurability with the Debye radius D and the Larmor radii ρ_{Hi} and ρ_{He}. The problems of such types that have so far being considered for different velocities of the bodies in the plasma ($V_0 \gg v_i$, $V_0 \sim v_i$ and $V_0 \ll v_i$) give us, as we shall see below, a fairly elegant description of the nature of the interaction between bodies and the plasma and of the effects that take place in the neighborhood of the bodies (Al'pert, Gurevich, Pitaevskii;[5,16] Al'pert, Gurevich, Pitaevskii, Smirnova;[18] Liu;[19] Singer (Editor);[21] Brundin (Editor);[20]). However, a serious shortcoming of the

present state of the theory is the absence of any completed investigation of even special character of nonstationary problems, when one takes into account the
terms $\partial f/\partial t$ in (1.11) and (1.12). Although the solu-
tions that are obtained are clearly related to the dispersion equation that describes the spectra of wave processes in the plasma (see §12), the excitation of waves
and the nature of the plasma instability in the neighborhood of a body have as yet been very little studied.

The fullest theoretical treatment has been given
for the following classes of problems.

1. In the "n e u t r a l" a p p r o x i m a t i o n, when
the paths of the ion motion are assumed to be rectilinear,
as is the case when neutral atoms and molecules encounter the body and are reflected from it (Gurevich[22]).
Naturally, in this case, the potential ϕ_0 of the body,
the electric field $E = \text{grad } \phi(r, \zeta)$ of the plasma, and
the geomagnetic field H_0 do not affect the motion of
the ions. Despite this considerable simplification, we
shall show below that in a certain i n t e r m e d i a t e
z o n e of distances, and under certain conditions near
the body as well, the results obtained in this approximation give a good description of the experimentally
observed perturbation of the density of charged particles behind the body if the potential of the body is
not too large.

2. With allowance for the influence of the e x t e r-
n a l m a g n e t i c f i e l d, when the charged particles
precess about H_0, but neglecting the influence of the
electric field of the plasma and the potential of the
body.[22] The corresponding theoretical results also agree
well with the results of measurements in a certain range
of distances.

3. In the f a r z o n e of the body at distances
$r \gg \rho V_0/v_i$ with allowance for the electric field, geomagnetic field, and also for the fact that the plasma
is nonisothermal, when the linearized problems were
solved (Pitaevskii;[23] Vas'kov;[24] Bud'ko[25]). This range
of distances has as yet been studied experimentally but
little, although the theoretical results are apparently
in good qualitative agreement with various experimental
data.

4. In the n e a r z o n e of the body with allowance
for the influence of the electric field for the case of

a weakly charged body of small diameter ($\rho_0 \ll D$) (Dub-ovoi[26]), and also an uncharged and a charged body of large diameter ($\rho_0 \gg D$) (Gurevich, Pitaevskii, Smirnova;[18] Moskalenko[27]). In these cases, the nonlinear problems were solved. The results of these investigations also agree fairly well with the results of experiments.

Experimental investigations have sometimes been made under conditions close to cases treated theoretically: by means of laboratory experiments or directly in the near-Earth plasma (on satellites and space probes). However, the majority of these experiments were not set up especially to match the requirements of theory, and their detailed theoretical analysis was not forseen. Therefore, it is generally possible to make only a qualitative comparison between the results of experiments and theory, although some of the data also agree well quantitatively. The difficulties that arise in the analysis of laboratory experiments are due primarily to the fact that the particle streams that encounter the model (body) were not sufficiently homogeneous over the cross section and had a large thermal spread; the extent to which they were nonisothermal (the value of the ratio T_e/T_i) were also unknown, nor did one know other important properties of the streams. It is therefore frequently impossible to compare experiments and theory in any accurate sense. The experiments made on artificial satellites and rockets are further complicated by the fact that these bodies frequently have a very complicated form; they have additional instruments that give rise to additional fluxes of reflected particles that encounter the measuring device (probe). As a rule, measurements were made near the surface of the satellite or rocket, when additional effects are most strongly expressed. At the same time, the fact that the ionosphere is a multicomponent plasma has an important but hard to estimate influence on the perturbation of the plasma. In addition, the fact that the plasma is nonisothermal plays a role, and this can be noted in a number of regions of the near-Earth plasma. Thus, if one is going to make a further and adequate theoretical analysis of the effect of the interaction between the plasma and the bodies, it is important to have experiments with bodies (satellites) of regular form (for example, a sphere) and to make simultaneous measurements of different plasma parameters. However, the literature does contain one series of laboratory experiments (Gurevich, Salimov, Buchel'nikova;[169] Bogashchenko, Gurevich, Salimov, Éidel'-man[28]) which were carried out in accordance with theor-

etical calculations made specially for the analysis of
the results of the measurements. This made it possible
to confront theory and experiment fairly accurately and
also quantitatively. Some of the data obtained in these
investigations are given below. This direction of laboratory investigations has been followed up particularly
strongly in recent times (Buchel'nikova, Éidel'man, et
al;[170] Schmitt[171]).

§9. Perturbations of the Plasma in the Neighborhood of
Rapidly Moving Bodies $(V_0 \gg v_i)$

Let us consider first supersonic flow of a stream
of neutral particles around a body, when the particles
are "swept aside" by the body, and an extended rarefied
wake is formed behind it.

1. Neutral Approximation. In this case, the particle paths are rectilinear. The electric and the magnetic field have no influence, and the problem reduces
to the solution of the kinetic equation (Gurevich[22])

$$v \partial f / \partial \mathbf{r} = A \delta(s), \tag{2.1}$$

which determines the distribution function $f_n(\mathbf{r}, \mathbf{v})$ and,
accordingly, the perturbed distribution of the particle
density $N_n(r, \zeta)$ in the neighborhood of the body. In
equation (2.1), $\mathbf{r}$ is the radius vector drawn from the
center of the body to the point of observation, and ζ
is the angle between $\mathbf{r}$ and an axis z that is collinear
with the direction of the velocity V_0 of the body. In
all that follows, the angle ζ will be measured to both
sides of the z axis, the positive direction of which is
in the opposite direction to that of the vector V_0, i.e.,
behind the body along the z axis we have $\zeta = 0$, while in
the forward direction $\zeta = \pi$; the value of z is measured
from the center of the body.

The distinctive feature of the perturbation in
this case is the accumulation of particles in front of
the body and the reduced density of particles behind it.
The accumulation occupies a small region and is appreciable only at distances from the body of the order of
its linear diameter. In contrast, the rarefied wake
of the body is very extended and is smeared out along
the z axis only at distances of the order of the mean
free path of the particles. At short distances from
the body, the particle density distribution depends

strongly on the form of the body. These features are
illustrated by the dependences of $N_n(z,0)/N_{n0}$ on z/ρ_0
for bodies of different form with the same cross section
s_0, which are shown in Figs. 7-9. Contours of equal values of $N_n(r,\zeta)/N_{n0}$ are plotted in Fig. 8 for a sphere[22]
and an ellipsoid (Sawchuk[29]) as obtained from the
solution of equation (2.1). These curves, like the ones
shown in Figs. 7 and 9, were calculated for $a_0 = V_0/v =$
8. In all that follows, the subscript 0 is appended to
the unperturbed particle densities. It can be seen from
these figures that near the body the particle density
falls very rapidly, approaching zero. The angular variation of the perturbed density is very strong at short
distances behind the body, as can be seen from Fig. 7,
in which the dependences $N_n(\zeta, r = \text{const})/N_{n0}$ are plotted
at the distances $r/\rho = 4$, 2, 1 for a sphere (continuous
curves) and an ellipsoid (dashed curves). At sufficiently
great distances from the body, namely, when

$$r/\rho_0 > V_0/v_i , \qquad (2.2)$$

$N_n(r,\zeta)$ already depends weakly on the form of the body
and is determined effectively by only the cross section
s_0 perpendicular to the vector $\mathbf{V}$. When (2.2) is satisfied, the perturbation of the particle density along the
z axis is fairly well described by the formula

$$\delta N = [N_n(z,0) - N_0]/N_0 = (s_0 a_0^2/\pi z^2) \exp(-a_0^2 \rho_0^2/z) , \qquad (2.3)$$

where $\rho_0 = \sqrt{s_0/\pi}$ is the effective radius of the maximal
cross section of the body. In the far zone, when

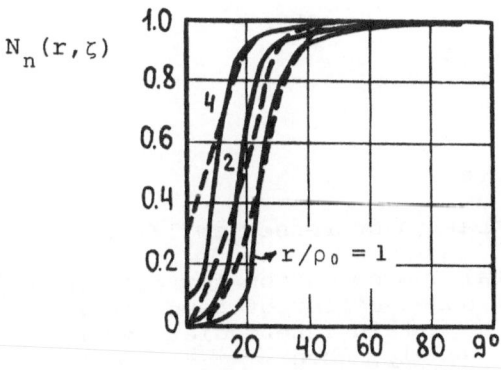

Fig. 7

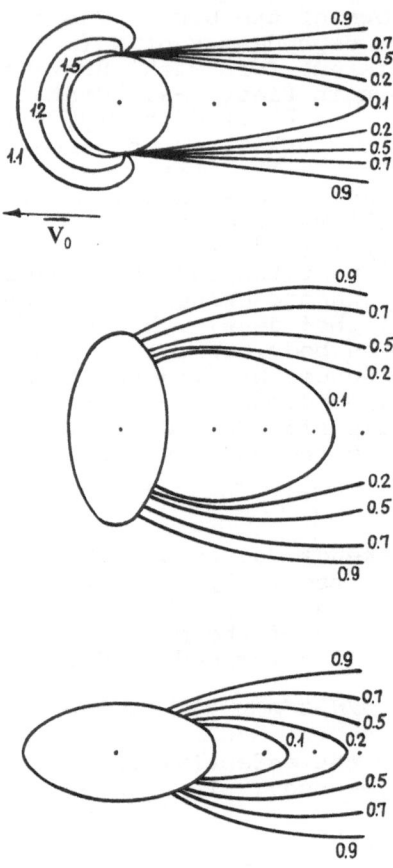

Fig. 8

$r/\rho_0 \gg V_0/v_i$,

$$\delta N \approx s_0 a_0^2/\pi z^2,$$ (2.4)

and the perturbation decreases as $1/z^2$.

The neutral approximation for determining the ion density $N(r, \zeta)$ under different conditions is evidently applicable in a certain intermediate zone of distances from the body, namely, when

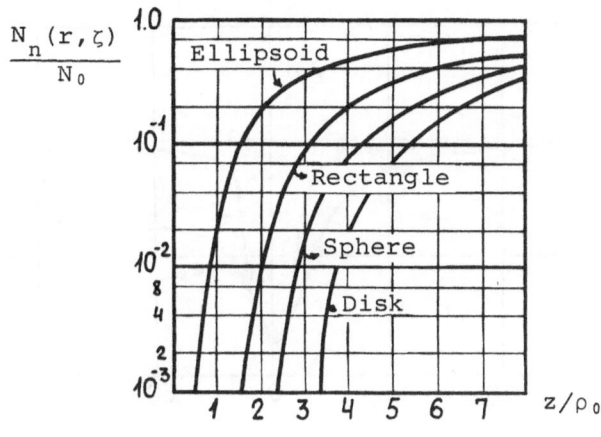

Fig. 9

$$\rho_0 V_0 / v_i \sim r \lesssim \rho_{Hi} V_0 / v_i. \tag{2.2a}$$

Nearer the body, the electric field begins to have an effect. At distances $r > \rho_{Hi} V_0 / v_i$, the geomagnetic field H_0 already has an appreciable effect, and the structure of the perturbation becomes quasiperiodic (see subsecton 2 below), while in the far zone, when $r \gg \rho_0 V_0 / v_i$, the influence of the electric field is particularly strongly manifested, and the angular dependence $N(r,\zeta)$ becomes more complicated. Naturally, the condition (2.2a) is meaningful only if $\rho_{Hi} > \rho_0$ or $\rho_{Hi} \gg \rho_0$, which is the case in the near-Earth plasma but is not always realized in laboratory experiments. In a nonisothermal plasma, when $T_e \gg T_i$, the influence of the electric field is stronger, and the applicability of the neutral approximation is more restricted.

Some results of experiments that illustrate the applicability of the theory of the "neutral" approximation for the determination of the angular dependence of the density of charged particles are given in Figs. 10-12. In Fig. 10 we have plotted the ratio $J_e/J_0 \approx N_e/N_0$, that is, the ratio of the changed electron flux J_e onto a probe placed at a distance $5\rho_0$ from the center of the satellite Ariel 1 to the unperturbed electron flux J_0

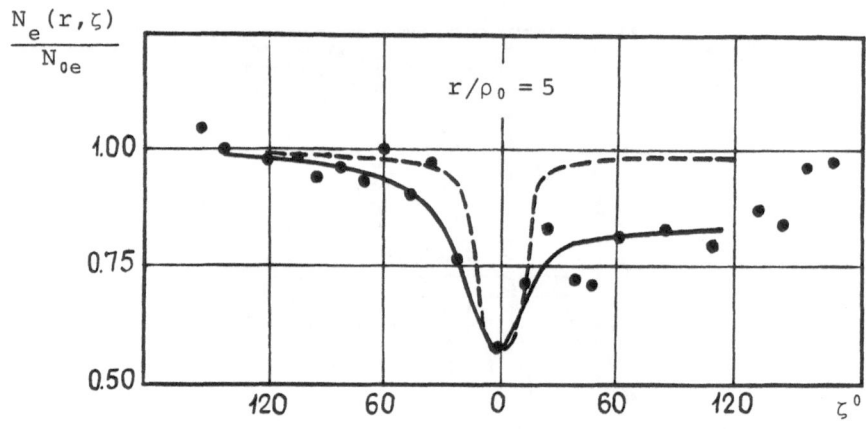

Fig. 10

(Bowen, Boyd, Henderson, Willmore;[30] Samir, Willmore;[31]
Henderson, Samir[32]). This satellite had a nearly spherical form, though it had a complicated structure of its
surface, and this is evidently the main factor responsible
for the large spread of the results of the measurements.
In a somewhat smoothed form, they are represented in
Fig. 10 (dots and heavy curve). The dashed curve shows
the results of a calculation of $N_n(r,\zeta)/N_0$ in the neutral
approximation for the mean value of $a_0 = V_0/v_i$; these
calculations corresponded to the conditions of the experiments (see ref. 18). Because of the appreciable
change of the ion composition and the temperature with
altitude, the value of a_0 changed appreciably in these
experiments, and this, in its turn, complicated the theoretical analysis of the results of the measurements.
In view of these circumstances, one can see that there
is satisfactory agreement between the results of the experiments and the calculations. An important point is
the almost exact agreement of the minimal values of
$N_e(r,\zeta)$ and $N_n(r,\zeta)$ for $\zeta = 0$.

The results of laboratory measurements (Clyden,
Hurdle[33]) of the angular dependence $\delta N_i(r,\zeta) = [N_i(r,\zeta) -
N_0]/N_0$ for $\phi_0 = 0$ and the corresponding theoretical curve
$\delta N_n(r,\zeta)$ are plotted in Fig. 11, and these also agree
fairly well. The theoretical dependence $\delta N_n(r,\zeta)$ is
plotted in Fig. 11 from the data of ref. 22. We should

Theory: $\phi_0 = 0$, $V_0/v_i = 8$, $\rho_0/D \gg 1$

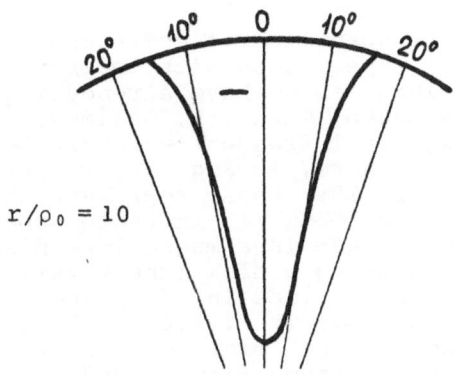

$r/\rho_0 = 10$

Experiment: $\phi_0 = 0$,
$V_0/v_s = 6\text{-}50$, $\rho_0/D \approx 5$

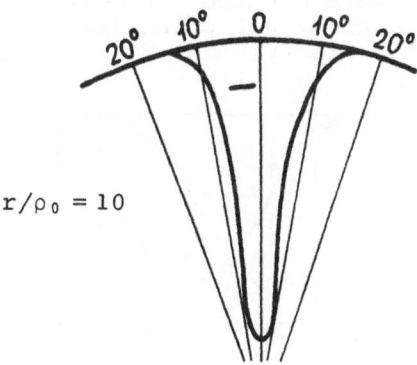

$r/\rho_0 = 10$

Fig. 11

point out here that in what follows we shall frequently, for the sake for a lucid representation, use a representation of the angular dependences as in Fig. 11. In these figures, the line in the form of an arc corresponds to the zero level $\delta N = 0$. Positive values, $\delta N > 0$, are plotted above the arc as a function of ζ; negative values, $\delta N < 0$, are plotted below the arc; the angle ζ is

measured behind the body to both sides of the direction
$-V_0$.

    A more accurate quantitative comparison of the re-
sults of laboratory experiments with theoretical calculations has been given, as we have already mentioned,
in ref. 28 (Bogashchenko, Gurevich, Salimov, Éidel'man).
In what follows, we shall frequently return to the results of this paper. Here, in Fig. 12, we have plotted
curves taken from ref. 28; these represent the dependences $N_i(y, z = const)/N_0$ obtained in these experiments.
The measurements were made in a magnetized plasma with
an ion stream encountering a disk (the y axis is perpendicular to the z axis and lies in the plane of the disk).
In the papers quoted above (ref. 170; Astrelin, Bogashchenko, Buchel'nikova, Éidel'man) results are also given
of an investigation of plasma flow around a plate, half-
plane, and cylinder, and experimental data are compared
with theoretical calculations in the same way as was
done in ref. 28. The curves in Fig. 12 correspond to
conditions of experiments in which $V_0/v_i \approx 2$ and ρ_{Hi}/ρ_0
≈ 1.5. The curves 1 were plotted from the results of
calculations in the neutral approximation; the curve 2
takes into account the electric field; and the dots give

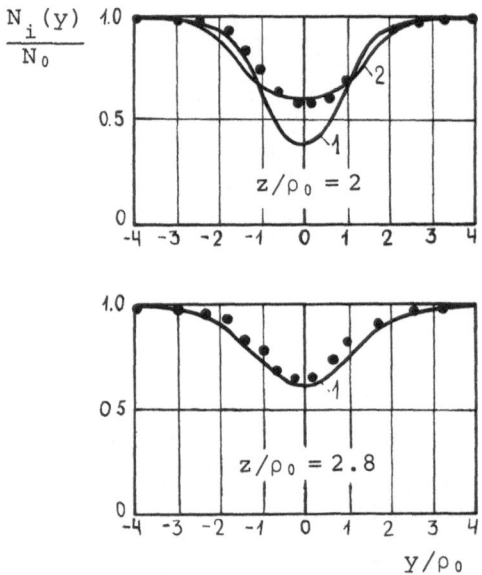

Fig. 12

the results of measurements. It can be seen that for
$z/\rho_0 = 2$ the neutral approximation gives a poor description of the results of the experiments in the shadow
region of the body, $y/\rho_0 \approx \pm 1$, while already at the distance $z/\rho_0 = 2.8$ the agreement is better at all values
of y. According to the data of these experiments, the
"neutral" approximation is applicable to $z/\rho_0 \sim 4-5$.
Subsequently, the magnetic field begins to have an influence, although the influence of the electric field
need only be taken into account when $z/\rho_0 \gtrsim 18-20$, i.e.,
when $z/\rho_0 \gtrsim (9-10)(V_0/v_i)$.

<u>2. Influence of the Geomagnetic Field</u>. If a geo-
magnetic field H_0 is present, the particles precess
about the vector H_0, and the paths of their motion become more complicated. It is however readily seen that
the influence of the magnetic field must be manifested
only at distances $z \gtrsim \rho_{Hi} V_0/v_i$, which will be evident
from what follows. For a body of circular cross sec-
tion moving along the magnetic field $(H_0 \| V_0)$ the ion den-
sity on the z axis behind the body $(\zeta = 0)$ can be well
described without allowance for the influence of the
potential of the body, the electric field, and collisions,
by means of the formula

$$N_i(z,0)/N_0 \approx \exp\{-\rho_0^2/[4\rho_H^2 \sin^2(\pi z/\Lambda_z)\}. \qquad (2.5)$$

(Gurevich[22]). It can be seen that the density of the
charges varies periodically along the z axis with spatial
period

$$\Lambda_z = 2\pi\rho_{Hi} V_0/v_i = 2\pi V_0/\Omega_{Hi}. \qquad (2.6)$$

Since the case $V_0 \| H_0$ is distinguished, in the absence
of collisions the perturbation δN_i of the density does
not decrease with the distance: for $\rho_{Hi} < \rho_0$ the charges
move in this case along spirals along H_0 and the rarefied
region -- the wake of the body -- is not filled with particles, having a cylindrical form. When collisions are
taken into account, the oscillatory nature of the wake
is gradually smeared out, and its structure becomes quasiperiodic, and the wake itself disappears at distances of
the order of the mean free path of the particles. It
is also easy to note that when

$$z/2\Lambda_z = z/2\rho_{Hi} a_0 \ll 1,$$
$$\sin(2\pi z/2\Lambda_z) \approx z/2\rho_{Hi} a_0 \qquad (2.7)$$

it follows from formula (2.7) that

$$\delta N_i \approx \rho_0 a_0^2 / z^2 = s_0 a_0^2 / \pi z^2. \tag{2.8}$$

Thus, in the near zone of the body, the perturbation of the ion density decreases as $1/z^2$ and the influence of the magnetic field disappears -- the formula of the neutral approximation applies and (2.8) agrees with (2.4). We have already pointed out this circumstance.

In the case of motion at right angles to the magnetic field ($V_0 \perp H_0$) the formulas for a rectangular plate have a simple form. Since there is no axial symmetry when $V_0 \perp H_0$, for the circular plates the formulas are very complicated. Along the z axis for a plate $s_0 = 4\rho_x \rho_y$

$$N_i(z,0)/N_0 = \exp \left\{ 1 - \Phi(a_0 \rho_x / z) \Phi[\tfrac{1}{2}\rho_y / \rho_{Hi} \sin(z/2a_0 \rho_{Hi})] \right\}$$

(see refs. 16 and 22), where
$$\tag{2.9}$$

$$\Phi(\alpha) = (2/\sqrt{\pi}) \int_0^\alpha \exp(-x^2)\,dx \tag{2.10}$$

is the error function. We see that when $V_0 \perp H_0$ the perturbation of the ion density also has a periodic structure, although in contrast to the case $V_0 \perp H_0$, the value of δN_i also decreases with the distance in the absence of collisions, although slower than in the neutral approximation, namely, on the average as $1/z$. In Fig. 13 we have plotted for comparison the dependences δN_i described by equations (2.3), (2.5), and (2.9).

A quasiperiodic structure of the wake of the body was first predicted theoretically[22] and then observed in ref. 34 (Barett). However, it was not until the experiments quoted above[28] that a quantitative comparison of theoretical calculations was made with the results of measurements with allowance for the effect of col-

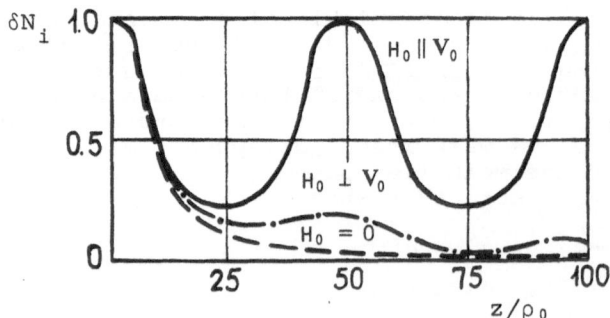

Fig. 13

lisions on the ion temperature. The results of two series of measurements of $J_i(0,z)/J_0 \approx N_i(0,z)/N_0$ are plotted in Fig. 14 for $V_0/v_i \approx 2$ and 2.6 and $\rho_{Hi} \approx 0.4$ and 0.2
cm, respectively. In the upper part of the figure one
can clearly see that there is complete agreement between
the experiment (points) and theory (continuous curves).
In the lower part of the figure, which corresponds to
measurements with a magnetic field that was approximately
twice as strong, the periodicity of the wake agrees well
with the theory, although the experimental values of
$N_i(0,z)$ are larger than the theoretical ones. This was
confirmed in these experiments in different series of
measurements. This nature of the discrepancy between
theory and experiment becomes more pronounced with increasing value of ρ_0/ρ_{Hi}. It is assumed in ref. 28
that since in a magnetic field the ion density along the
z axis increases, which is due to the effect of the electric field (Pitaevskii,[23] Vas'kov[24]), in the calculations it is necessary, beginning with certain values of
H_0, to take into account the influence of the electric
field. With increasing ρ_0/ρ_{Hi}, the wake of the body
becomes more rarefied, and the role of the electric field
increases.

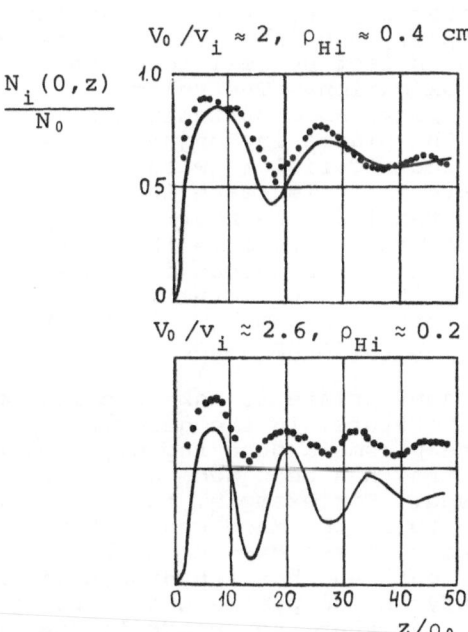

Fig. 14

   3. Influence of the Electric Field. An electric
field $E(r,\zeta)$ arises in the plasma in the neighborhood
of a moving body because the ions and electrons follow
different paths in their motion. This leads to a difference in their densities: $N_e(r,\zeta) \neq N_i(r,\zeta)$. It is
also obvious that, in addition, the charge of the body
also creates an electric field, and it is this source
of the field E that in a number of cases is decisive in
the formation of the structure of the plasma perturbation. Naturally, the electric field must play an important role out to the boundary of the region of Debye
shielding. At the same time, the influence of the electric field is also large in the far zone of the body,
where the plasma is quasineutral: $N_e \approx N_i$. This is because at large distances, due to the small perturbation
δN of the particle density, the relative influence of
the field increases and, as we shall see below, the
perturbation of the plasma has a complicated structure
-- the angular dependences $N_i(r,\zeta)$ are more complicated
than near the body. In addition, far from the body,
the role of the geomagnetic field is greater, as we have
seen. Conversely, in the immediate proximity of the
body and behind it in the region in which the plasma is
most strongly rarefied, the structure of the perturbation becomes simpler, and in a number of cases the perturbation δN can in fact be well described by means of
the neutral approximation. The relative size of the
body, ρ_0/D, also plays an important role in forming the
perturbation of the plasma in conjunction with the influence of the electric field; with increasing diameter
of the body, there is an increase in the relative influence of its potential, and near the body one has effects similar to those observed in the case of high potentials of a large body at great distances from the body.
At the same time, it must be borne in mind that a nonisothermal state of the plasma ($T_e \neq T_i$) also plays an
appreciable role in these processes.

   It is here appropriate to make the following com-
ment. In the literature, in the description of effects
that arise in the plasma around rapidly moving bodies,
the terminology that has been adopted in gas dynamics
is frequently used. For example, the wake behind the
body is referred to as the Mach cone. However, even the
simple analogy between the Mach cone that is observed
in a "continuous gas" and has a hydrodynamic nature and
the wake of a body in the plasma has a purely formal
nature: one can say that the anlogy is "geometrical".
In contrast to the Mach cone, which has abrupt boundaries,

the structure of the wake of a body in the plasma is of
kinetic origin and depends strongly on the influence of
the electric field. Its boundaries are generally smoothed
out, this being due to the influence of the damping
of ion-acoustic waves. In a number of cases the wake
has a multipetal structure, regions of depletion and
accumulation of particles alternating, etc. Thus, both
as regards its nature and its structure, the wake of a
body in the plasma is very different from a Mach cone.
We may also point out that in gas dynamics a characteristic feature is the formation of shock waves in front
of a body, which does not occur in the phenomena that
we are considering. There is therefore neither point
nor any physical justification for using here the hydrodynamic terminology and referring to the wake of a rapidly moving body in the plasma as the Mach cone. I believe
that such terminology leads to confusion in the physical
understanding of the different types of phenomena.

Before we turn to a more detailed description of
the phenomena considered in this section, let us point
out their general and fundamental features.

1. In the immediate proximity of the surface of the
body, under the influence of the electric field, the
density of charged particles is appreciably higher than
that expected in the neutral approximation. The angular dependence $N(r,\zeta)$ agrees qualitatively with the expression $N_n(r,\zeta)$ for neutral particles.

2. Under different conditions (see below) charged
particles are focused behind the body. The region in
which the plasma is most strongly rarefied lies to both
sides of the axis on a conical surface with opening
angle $\zeta_m \sim \sin^{-1}(v_i/V_0)$ or $\sim\sin^{-1}(v_s/V_0)$ ($v_s = \sqrt{\kappa T/M}$ is
the velocity of nonisothermal sound).

3. The focusing in the neighborhood of the axis of
the body becomes so strong in a number of cases that the
density $N_i(r,\zeta)$ is higher in a certain sector of angles
$\Delta\zeta$ than the density N_0 of the unperturbed plasma, i.e.,
$\delta N > 0$, and a region in which particles accumulate is
formed.

4. The focusing effects mentioned under headings
2 and 3 above are due to the influence of the potential
of the body and the nonisothermal state of the plasma,
and for small bodies they become stronger. In various
experiments it has been established that $\delta N > 0$ already

for values of the potential of $\phi \sim -(1-2)V$.

5. As one moves away from the body, at sufficiently great distances from it, one can find that two regions of accumulation arise behind the body, not on its axis, but to the side of the axis. In this case, one observes one or three depletion regions.

6. These effects are possible both in the presence and the absence of a geomagnetic field H_0. Under the influence of a magnetic field, the structure of the perturbation is smoothed out in the far zone and becomes asymmetric with respect to the axis of rotation if the velocity vector V_0 makes with H_0 an angle $\theta_0 \neq \pi/2$.

4. Near Zone: <u>Large Body</u> ($\rho_0 \gg D$; $r \lesssim \rho_0 V_0/v_i$, $r \ll \rho_0 V_0/v_i$) <u>and Small Body</u> ($\rho_0 \ll D$; $r \lesssim DV_0/v_i$, $r \ll DV_0/v_i$). The theoretical formulas that take into account the influence of the electric field are very complicated and are usually expressed in terms of integral expressions. Therefore, the results of the solution of different problems can be comprehended only by means of numerical calculations. In a number of cases one employs a numerical solution of the differential equations of motion of the particles and the Poisson equation, and the results are not represented analytically but only in graphical form. However, for some special cases one can nevertheless obtain fairly simple formulas. This enables one to compare directly the results of experiments with the theory if the conditions of the experiments correspond to the restrictions of the theory or, conversely, to arrange experiments that satisfy conditions met by the formulas.

For an infinitely long circular cylinder whose radius satisfies $\rho_0 \gg D$ moving in an isothermal plasma in the direction normal to its axis, the perturbed value of the density in the region near the surface of the cylinder has a fairly simple form (Gurevich, Pitaevskii, Smirnova[18]):

$$N(r,\zeta)/N_0 = A(T_e/T_i)\{\exp[-(V_0/v_s)X - \tfrac{1}{2}X^2]$$
$$+ \alpha \exp[-(V_0/v_s)X - \tfrac{1}{2}X^2]\}, \qquad (2.11)$$

$$X = \pi - \zeta - \sin^{-1}(\rho_0/r),$$

where $A(T_e/T_i)$ varies in the range from 0.7 to 0.4 as T_e/T_i varies from 1 to ∞, and $\alpha = 1$ or 0 depending on whether or not the radius vector r makes with the direction $-V_0$ an angle

$$\zeta > \pi/2 - \sin^{-1}(\rho_0/r) \quad \text{or} \quad \zeta < \pi/2 - \sin^{-1}(\rho_0/r),$$

i.e., depending on whether or not the point of observation lies above or below the shadow of the cylinder -- the tangents to the section of the cylinder parallel to the vector V_0.

On the axis ($\zeta = 0$) behind a circular disk of radius $\rho_0 \gg D$ (see ref. 18),

$$N(z,0)/N_0 = A(T_e/T_i)\sqrt{2\pi}\rho_0(z^2 + \rho_0^2)^{-\frac{1}{2}}(T_e/T_i)$$
$$\times \{V_0/v_s + z/\rho_0 + \tan^{-1}(\rho_0/z) \tag{2.12}$$
$$\times \exp[-(V_0/v_s)\cot^{-1}(\rho_0/z) - \tfrac{1}{2}\arctan^2(\rho_0/z)]\},$$

where $A(T_e/T_i)$ varies in the same range as in formula (2.11).

Simple formulas are also obtained for an isothermal plasma and weakly charged bodies of small diameter ($\rho_0 \ll D$, strictly speaking for point charges $Q \approx \phi_0\rho_0$, $Q < 0$) when the following conditions hold:

$$B = (c|Q|v_i)/(\kappa T V_0 D) = (e|\phi_0|\rho_0 v_i)/(\kappa T V_0 D) \ll 1,$$
$$\frac{e|\rho_0|}{\kappa T} \lesssim V_0/v_i. \tag{2.13}$$

In this case (Dubovoi[26]) at distances satisfying

$$r/D \ll V_0/v_i \quad \text{or} \quad r/D \lesssim V_0/v_i, \tag{2.14}$$

the perturbation δN_i of the ion density is

$$\delta N_i(r) = [N_i(r,\zeta) - N_0]/N_0$$
$$= (e|\phi_0|\rho_0/\kappa T\iota) F(V_0/v_i, \sin\zeta) \tag{2.15}$$

and asymptotically, behind the body,

$$F(V_0/v_i, \sin\zeta) = -2.2\{I_m[(\frac{V_0}{v_i}\sin\zeta)^2 + 0.6 + 0.87i]^{-3/2}. \tag{2.16}$$

In (2.16), I_m is the imaginary part of the expression in the square brackets. The field within the region of Debye shielding ($r \ll D$) is a Coulomb field, i.e.,

$$\phi(r) = |Q|/r = -|\phi_0|\rho_0/r, \qquad (2.17)$$

and accordingly the perturbation of the electron density is

$$\delta N_e(r) = -e|\phi_0|\rho_0/\kappa T_e r. \qquad (2.18)$$

At greater distances, namely, for $r \lesssim D$,

$$\phi(r) = -|\phi_0|(\rho_0/r)\exp(-r/D),$$
$$\delta N_e(r) = -e|\phi(r)|/\kappa T \qquad (2.19)$$

The dependence $N(r,\zeta)/\kappa T$ calculated in accordance with formula (2.12) is shown in Fig. 15, where it is compared with the results of measurements[20-32] of the ratios of the electron fluxes on the surface of the satellite Ariel 1, which had an approximately spherical form. In the calculations, a mean value of V_0/v_i satisfying the conditions of the experiments was chosen. The dashed section in the theoretical curve corresponds to the interval of angles $\zeta = \pm 60°$, where the theoretical formula is less valid. Generally, one observes a fairly good agreement not only qualitatively but also quanti-

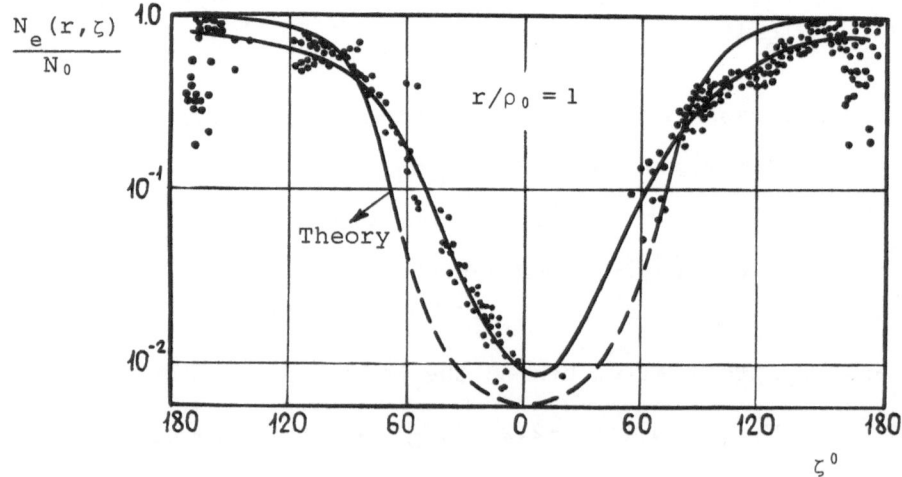

Fig. 15

tatively between theory and experiment. It should be
noted that the experimentally measured minimal value
$N_e(r,0)/N_0 \sim 10^{-2}$ at $\zeta = 0$ is, in close agreement with
theory, larger by approximately three orders of magnitude than the corresponding value $N_n(r,0)/N_0 \sim 10^{-5}$
obtained under the same conditions in accordance with
the formulas of the neutral approximation. This, as we
have already mentioned above, demonstrates the considerable influence of the electric field near the body.

Similar results of confrontation of theory with
experimental data (see ref. 18) are shown in Fig. 16,
in which the results of the measurements correspond to
experiments on the satellite Explorer 31 (Samir, Wrenn[35]),
which had the form of an octahedral parallelepiped, in
fact, nearly a cylinder. The upper part of the figure
corresponds to conditions when the measured electron
density was near the ion density, since there was a
single predominant species of ions (oxygen ions, O_1^+) in
the ionosphere and $V_0/v_i = V_0/v_i(O_1^+) \approx 5$. In this case,
formula (2.12) works well. As can be seen from the figure, good agreement is obtained between experiment and
theory, as in Fig. 15. However, in the same experiments
but under other conditions protons (H_1^+) were predominant
in the ionospheric composition, and for these $a_0 = V_0/v_i$
(H_1^+) ~ 1.2, i.e., a_0 was smaller by a factor of approximately 4 than for the measurements whose results are
shown in the upper part of Fig. 16, when the relative
content of O_1^+ was 99%. If the H_1^+ constituent is predominant the influence of the electric field on the
motion of the ions is greatly reduced because of the
smaller value of a_0, and this influence can be ignored
in the analysis of results of measurements at the boundary of maximal depletion, where the values of N_e and N_i
become approximately equal. In ref. 18 this boundary
was characterized by the angle α it subtended at the points
at which the corresponding probes were situated. For
these experiments, Gurevich, Pitaevskii, and Smirnova[18]
chose the angle $\alpha \approx 45°$; evidently, this value of α
may be universal in may experiments. Then, in the given
cases, when the relative proportion of protons was higher than 30%, the following approximate formula of the
neutral approximation, which determines the plasma density near the body, is recommended in ref. 18 for the
conditions specified above:

$$N(\zeta)/N_0 \approx [N(O_1^+)/N_0 + 1 + \Phi(V_0/v_i(O_1^+) - \cos\phi_0\cos\zeta)]$$
$$+ [N(H_1^+)/N_0 + 1 + \Phi(V_0/v_i(H_1^+) - \cos\phi_0\cos\zeta)],$$

$$(2.20)$$

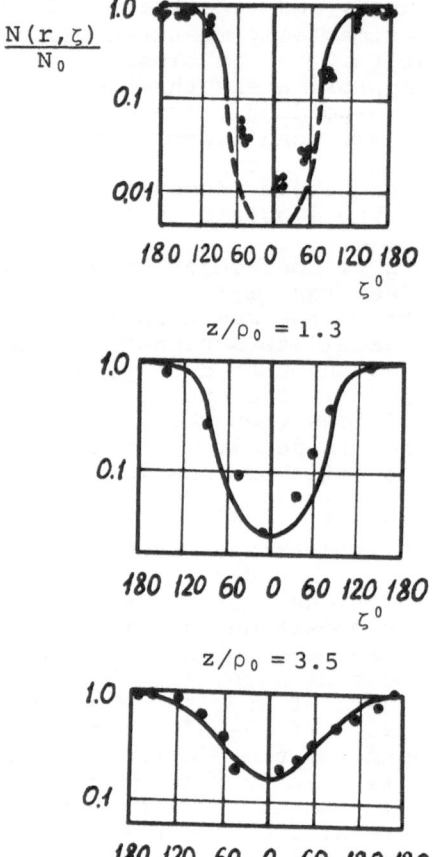

$$z/\rho_0 = 1.3$$

$$z/\rho_0 = 3.5$$

Fig. 16

where $\Phi(\ )$ is the error function (2.10). In the lower
part of Fig. 16 we show two cases of comparison of the
results of measurements (points) with theoretical curves
calculated in accordance with (2.20) for relative values
of H_1^+ equal to 0.23 and 0.94 at distances from the center
of the body of $z/\rho_0 \approx 1.3$ and 3.5. One can see that there
is good agreement between formula (2.20) and the results
of the measurements even at a fairly large distance from
the body, when the protons were predominant and $V_0/v_i(H_1^+)$
~ 1.2. Recently, an investigation has been made of the

influence of an electric field on the motion of ions in a multicomponent plasma (Gurevich, Pariiskaya, Pitaevskii[175]) and good agreement with formula (2.20) of the quasineutral approximation was established.

In a nonisothermal plasma in the near zone of the body, the effect of focusing of the particles around the axis of the body increases gradually as one moves away from its surface. The corresponding theoretical dependence $\delta N_i(\zeta)$ for $T_e/T_i = 4$ and $r/\rho_0 = 4.5$, shown in Fig. 17, was constructed from the data of ref. 18. It follows from Fig. 17 that for $\zeta = 0$ the ions are depleted by about twice as much as they are to the side (the dashed curve is for $T_e = T_i$). The experimental dependence $\delta N_i(\zeta)$, which is given in the upper part of Fig. 18 in in accordance with the data of ref. 36 (Skvortsov, Nosachev), illustrates the same effect and is in good qualitative agreement with the theoretical curve of Fig. 17. According to the estimate made by Bud'ko[37] $T_e/T_i \gtrsim 5$ in these measurements. In the lower part of Fig. 18 we have plotted, again from the results of these experiments for a different case,[36] the angular dependence $\delta N_i(r,\zeta)$, which shows that the particles were so strongly focused that $\delta N_i > 0$, and the perturbed value of the ion density was such that $N_i(r,\zeta) > N_0$. Positive perturbations, δN_i

$$\phi_0 = 0, \quad V_0/v_i \approx 8, \quad \rho_0/D \gg 1$$

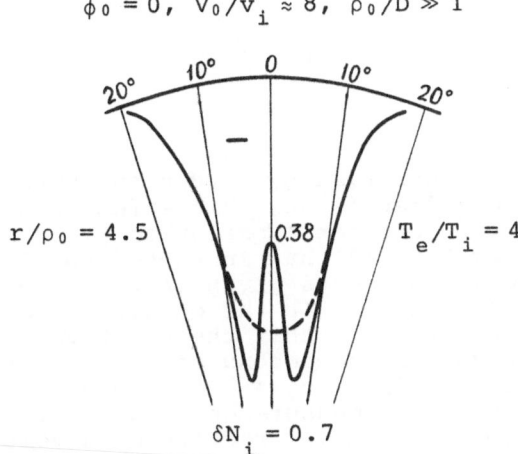

Fig. 17

$\phi_0 \sim 0, \quad V_0/v_s \approx 11, \quad \rho_0/D = 30, \quad T_e/T_i \approx 5$

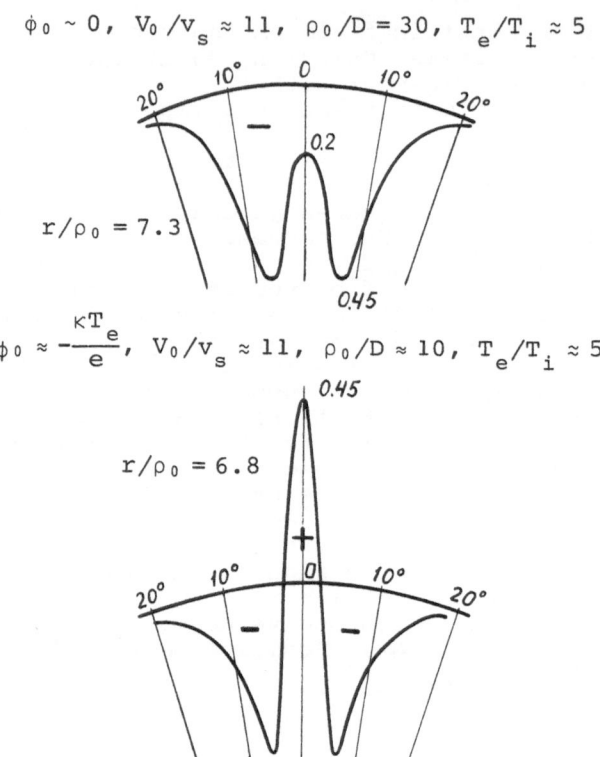

$\phi_0 \approx -\dfrac{\kappa T_e}{e}, \quad V_0/v_s \approx 11, \quad \rho_0/D \approx 10, \quad T_e/T_i \approx 5$

Fig. 18

> 0, were already observed by Clyden and Hurdle[33] and Hall, Kemp, and Sellen,[38] and this effect was predicted theoretically even earlier (Mosalenko[39]). However, it can be seen from Fig. 18 that in these experiments the results of the measurements given in the upper and lower part of the figure differ in the conditions of the experiment only by the fact that the relative diameter of the body, ρ_0/D, in the second case was only a third of its value in the first case; the potential of the body in both series of measurements was very low, since $\kappa T_e/e$ was only a very small fraction of a volt ($\sim 10^{-2}$, 10^{-3} V). It is possible that it is the relative decrease in the diameter of the body that explains the enhancement

of the focusing effect in these experiments. Indeed,
it has been shown theoretically for very small bodies
($\rho_0 \ll D$) that the focusing effect is enhanced (Dubovoi[26]).
Thus, at distances $r \sim D$ there is no depletion of particles at all ($\delta N > 0$). It is only with increasing distance, as can be seen from Fig. 19, that depletion regions gradually appear, these becoming appreciable at
large distances from the body (see also the following
subsection). In one of the recent experimental investigations (Hester, Sonin[40]) the evolution of the angular
dependence $\delta N_i(r,\zeta)$ was followed in detail behind a
sphere of two diameters: $\rho_0/D = 14$ and $\rho_0 = 1.8$. The
potential of the bodies in these experiments was only
fractions of a volt. Some results of these measurements

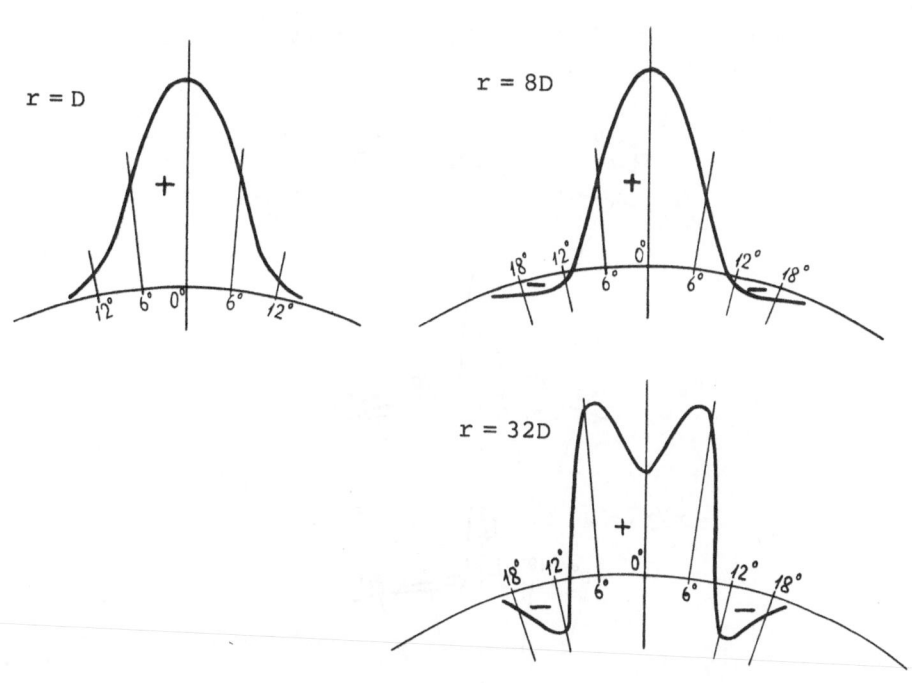

$\rho_0 \ll D, \quad V_0/v_i = 8, \quad T_e/T_i = 1$

Fig. 19

are given in Figs. 20 and 21. They show that distances
$r/\rho_0 \sim V_0/v_s$ from the body one already observes, as in
Fig. 19, a positive focusing of the ions ($\delta N_i > 0$). At
the same time, for $r/\rho_0 > V_0/v_s$ the angular dependences
$\delta N_i(r, \zeta)$ become more complicated and are similar to the
corresponding theoretical dependences, which are considered in the following subsection for the far zone of
the body, $r/p_0 \gg V_0/v_i$.

This focusing effect naturally leads to a complicated distribution of the electric field of the plasma
around the body. Corresponding results of calculations
for a large body ($\rho_0/D = 50$) are shown in Fig. 22, on
which we have plotted contours of equal potential (expressed next to the curves in relative units) behind
the body obtained from a selfconsistent solution of the
kinetic equation and the Poisson equation (Liu, Jew[41]).
Around the z axis, in the region of distances $r/\rho_0 \sim 2.5$
the potential of the plasma is maximal. The dashed
curves in the same figure show equipotential lines obtained

$$e\phi_0/\kappa T = -3.5, \quad V_0/v_s = 10.5, \quad \rho_0/D = 14, \quad T_e \gg T_i$$

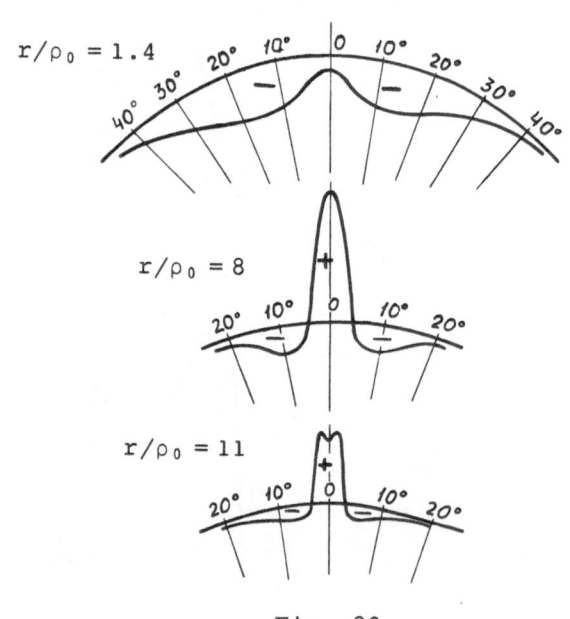

Fig. 20

$$e\phi_0/\kappa T = -3.5, \quad V_0/v_s = 8, \quad \rho_0/D = 1.8, \quad T_e \gg T_i$$

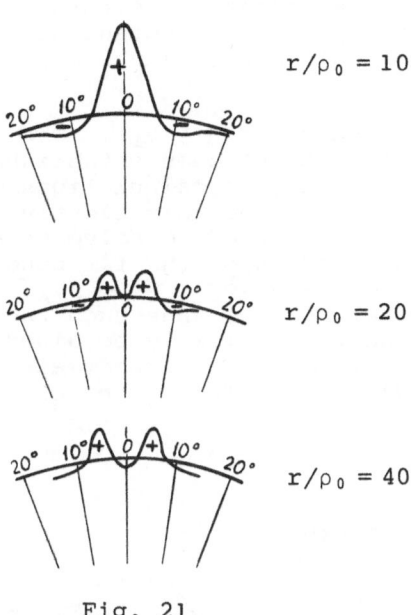

$r/\rho_0 = 10$

$r/\rho_0 = 20$

$r/\rho_0 = 40$

Fig. 21

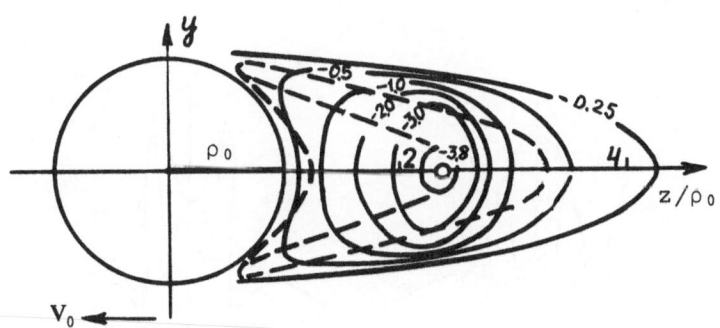

Fig. 22

from approximate formulas (see ref. 5). Near a small
body ($\rho_0 \ll D$) the angular variation of the potential
also exhibits a number of features (see Fig. 23). In a
nonisothermal plasma the potential of the plasma behind
the body also has positive value in the near zone (Dubovoi[26]).

    5. Far Zone ($r \gg \rho_0 V_0/v_i$). The far zone of the wake
of a body has not yet been established, since there are
no results of detailed calculations or experiments that
would enable one to do this clearly. Theoretically,
this is due to the need to solve equations of nonliinear
type. When speaking of the far zone one generally means
distances $r \gg \rho_0 V_0/v_i$ or $r \gg \rho_0 V_0/v_s$. However, it will be
seen from what follows that when these conditions are
satisfied the theoretically obtained effects have been
observed experimentally already at distances $r \lesssim \rho_0 V_0/v_i$
(see Fig. 21), while for bodies of small diameter ($\rho_0 \ll$

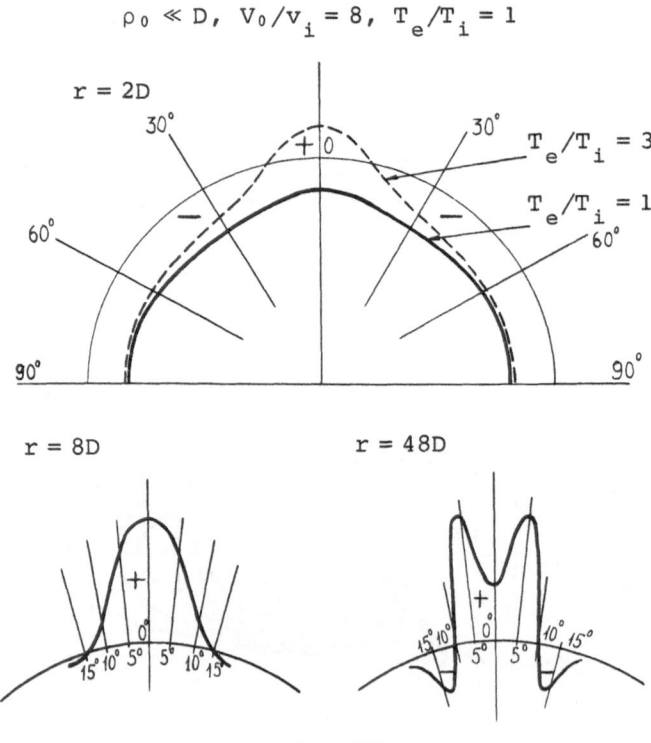

$$\rho_0 \ll D, \quad V_0/v_i = 8, \quad T_e/T_i = 1$$

Fig. 23

D) they are already manifested at distances $r \sim (5-7)D$.
Thus, the division into zones is conditional, depending
strongly on the relative diameter ρ_0/D of the body, the
extent to which the plasma is nonisothermal (T_e/T_i),
and the relative value of the potential of the body,
$e\phi_0/kT$. The theoretical results given in this section
were however obtained under conditions when the above
far-zone condition holds. However, the elucidation of
the meaning of the symbol $\gg$ in this inequality requires
further investigations.

The main feature of the angular distribution of
the density in the far zone of a large body is the focusing of electrons and ions in the neighborhood of the
direction $-V_0$ along the axis behind the body (Panchenko,
Pitaevskii;[42] Bud'ko;[25] see Fig. 24). The corresponding
angular dependences of δN for different values of V_0/v_i
are shown in Fig. 25a for a sphere in the absence of a
magnetic field ($H_0 = 0$, Panchenko,[43]), for a cylinder
(Fig. 25b) for $H_0 = 0$ (Panchenko,[43] Vas'kov[24]), and for
the sphere when $H \neq 0$ and $V_0/v_i = 8$ for different values
of the angle θ_0 between the magnetic field H_0 and V_0
(Fig. 25c, Vas'kov[24]). A magnetic field weakens the
focusing of the particles and for angles $\theta_0 \neq \pi/2$ the
distribution is asymmetric with respect to the direction
of the vector V_0. It should be noted that when $|\phi_0| \ll$
$(\rho_0/D)^{4/3}kT/e$ the influence of the potential of the body
is generally small, and it can be ignored.

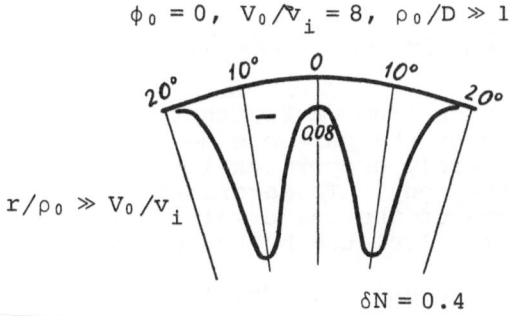

$$\phi_0 = 0, \quad V_0/v_i = 8, \quad \rho_0/D \gg 1$$

$$r/\rho_0 \gg V_0/v_i$$

$$\delta N = 0.4$$

Fig. 24

a. Sphere ($H_0 = 0$)

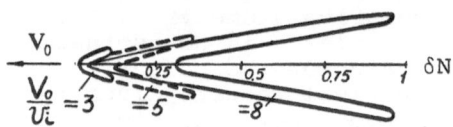

b. Cylinder ($H_0 = 0$)

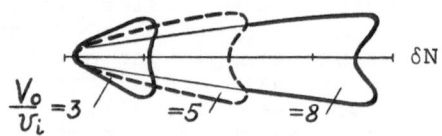

c. Sphere ($H_0 \neq 0$), $V_0/v_i = 8$

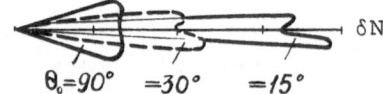

Fig. 25

Analytically, for a sphere $\rho_0 \gg D$ the perturbation of the particle density in a nonisothermal plasma is

$$\delta N(\mathbf{r}) = (\rho_0^2 V_0^2/r^2 v_i^2) B_0 (\zeta, V_0/v_i) . \qquad (2.21)$$

It can be seen from (2.21) that in the far zone the perturbation of the particle density decreases as $1/r^2$, i.e., inversely proportional to the square of the distance from the body. To determine $B_0(\zeta, V_0/v_i)$ one can use the graph of the universal function $F_0((V_0/v_i) \sin \zeta)$ (Fig. 26) calculated by Bud'ko,[25,37] where

$$B_0 (\zeta, V_0/v_i) = \cos \zeta \cdot F_0 \left((V_0/v_i) \sin \zeta \right) . \qquad (2.22)$$

For a nonisothermal plasma, the universal function also depends on the ratio T_e/T_i and it cannot therefore

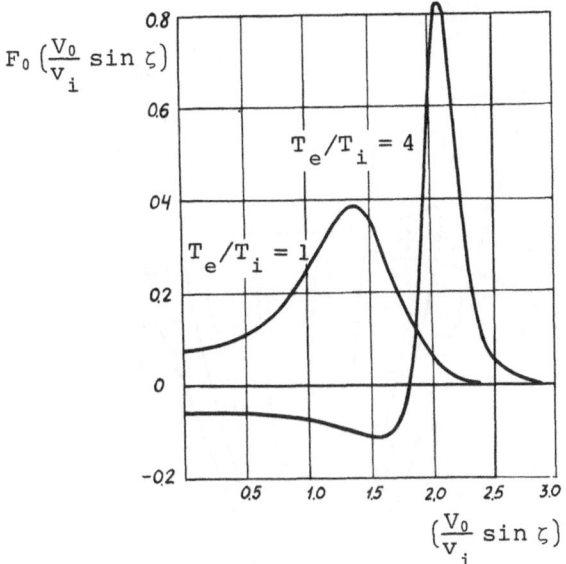

Fig. 26

be shown by a single graph. In Fig. 26 we have plotted
the universal function for $T_e/T_i = 4$, by means of which
we determined the angular dependence of the perturbation
$\delta N(\zeta)$ of the density shown in Fig. 27. A characteristic
feature of the influence of the nonisothermal state of
the plasma in the far zone is, as can be seen from Figs.
26 and 27 and the results of ref. 37, the enhancement
of the focusing in the neighborhood of the direction of
motion of the body; beginning with $T_e/T_i = 1.76$, this
leads to the appearance of a region of positive values
of perturbation, $\delta N(\zeta) > 0$. Witn increasing T_e/T_i, the
structure of the perturbation acquires "spikes" (see
Fig. 28), which apparently "anticipate" the appearance
in a collisionless plasma of a "shock wave" type phenomenon. However, this question requires further investigation.

We should point out here that when T_e/T_i decreases
and becomes less than unity the influence of the electric
field is appreciably weakened. When $T_e/T_i \leqslant 0.23$, the
maximal depletion is already established on the axis
behind the body as in the case of neutral particles and
in the limit $T_e/T_i \to 0$ the corresponding formulas for δN

$$\phi_0 = 0, \quad V_0/v_i = 8, \quad \rho_0/D \gg 1, \quad T_e/T_i = 4$$

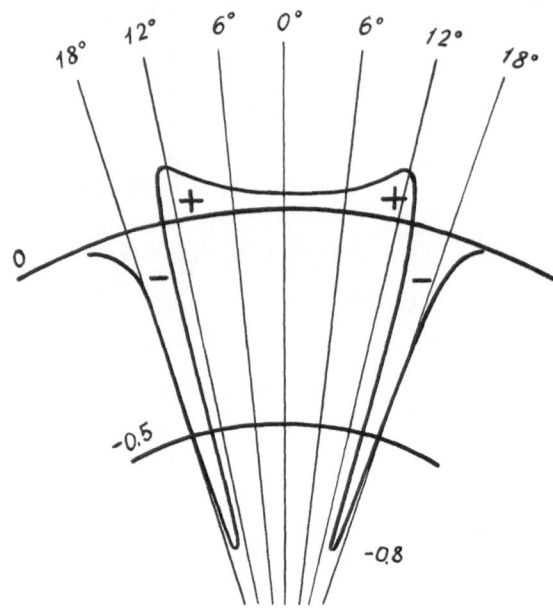

Fig. 27

go over into the formulas of the neutral approximation.

The wake of a cylindrical body of infinite length for which the projection of the section onto the plane perpendicular to V_0 has length equal to $2\rho_0$ is described by

$$\delta N(r) = -(2\rho_0/r)(V_0/v_i)B_{\parallel}\left((V_0/v_i)\sin \zeta\right) \qquad (2.23)$$

(Panchenko,[42] Bud'ko[37]). In contrast to $\delta N(r)$ for the wake of a sphere, $\delta N(r)$ in the far wake of a cylinder decreases as $1/r$. The dependences of the function $B_{\parallel}\left((V_0/v_i)\sin \zeta\right)$ are shown for $T_e/T_i = 1$ and $T_e/T_i = 4$ in Fig. 29.[36] As can be seen from Fig. 29, the wake of the cylinder is depleted of particles in the whole interval of angles ($\delta N < 0$, $B_{\parallel} > 0$). This is because the wake of a cylinder of infinite length is filled with particles only from its lateral surface. Around a cylinder of

$$\phi_0 = 0, \quad V_0/v_i = 10, \quad T_e/T_i = 32, \quad \rho_0/D \gg 1.$$

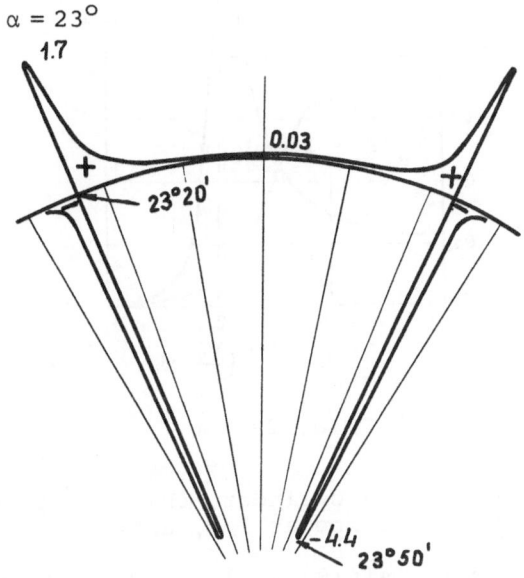

$\alpha = 23^\circ$

Fig. 28

of finite length in a nonisothermal plasma there must
be a region of accumulation of particles, as around a
sphere. However, the corresponding calculations have
not yet been made. In this case, the problem becomes
two-dimensional, and it is difficult to take into account the boundary conditions on the bases of the cylinder.

With allowance for the influence of an geomagnetic
field ($H_0 \neq 0$), the perturbation of the electron density
in an isothermal plasma behind a large body whose cross
section perpendicular to the vector $\vec{V}_0$ is $s = \pi\rho_0^2$ (ρ_0 is
the effective radius of the body) with averaging of δN
over the direction perpendicular to H_0 is also determined
by the following simple expression (Vas'kov[24]):

$$\delta N(\mathbf{r}, \zeta) = -\frac{\pi\rho_0^2 V_0}{\nu\rho_{Hi}v_i}B_H(\zeta, \theta_0, V/v_i). \qquad (2.24)$$

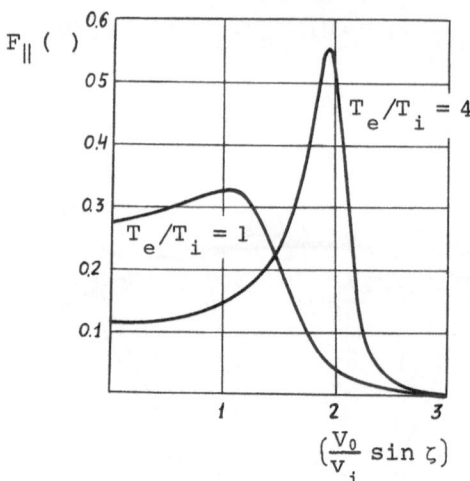

Fig. 29

As in the case of a cylinder, $\delta N(r,\zeta)$ decreases in this case with the distance as $1/r$, and

$$B_H(\zeta,\theta_0,V_0/v_i) = [\sin(\theta_0 - \zeta)]^{-1}F[V_0 \sin \zeta/v_i \sin (\theta_0 - \zeta)].$$

$$(2.25)$$

where θ_0 is the angle between the vectors $\mathbf{V_0}$ and $\mathbf{H_0}$. The form of the universal function $F(\ )$ in (2.25) is the same as that of the universal function $B_{\parallel}(\ )$ for a cylinder in an isotropic plasma, which is shown graphically in Fig. 29. Since the perturbation is over small angles $\pm\zeta$, we can replace $\sin \zeta$ by ζ in (2.25). It should be borne in mind that formula (2.25) applies only when $\sin(\theta_0 - \zeta) \geqslant 0$. If $\sin(\theta_0 - \zeta) < 0$, then $F_H(\ ) = 0$.

The wake of a long cylinder in the presence of a magnetic field is described by the same universal formula (Vas'kov[24]), namely,

$$\delta N(r ,\theta) = -(2\rho_0 V_0 /rv_i)B_H(\zeta,\theta,V_0 /v_i) \qquad (2.26)$$

where, as above (see (2.23)), $2\rho_0$ is the length of the projection of the section of the cylinder onto the plane perpendicular to $\mathbf{V_0}$.

In the far zone of the body, as we have already mentioned several times above, the plasma is quasineutral.

Therefore, the potential of the plasma is proportional to the perturbation of the electron density, i.e.,

$$\phi = (\kappa T_e/e)\,\delta N, \tag{2.27}$$

and, for example, for a sphere using (2.21), one can obtain the following expressions for the radial and the angular component of the electric field in an isothermal plasma:

$$E_r = -(2\kappa T V_0^2 \rho_0^2/ev_i^2 r^3)\,B_0(\zeta, V_0/v_i),$$
$$E_\zeta = (\kappa T V_0^2 \rho_0^2/ev_i^2 r^3)\,B_{0\zeta}'(\zeta, V_0/v_i), \tag{2.28}$$

where the derivative of $B_0(\ )$ with respect to ζ is equal to

$$B_{0\zeta}'(\zeta, V_0/v_i) = -\sin\zeta\, F_0\left(\frac{V_0}{v_i}\sin\zeta\right) + \frac{V_0^2}{v_i^2}\cos^2\zeta\, F'\left(\frac{V_0}{v_i}\sin\zeta\right)$$

$$\tag{2.29}$$

The curves of $F_0(\ )$ and $F_0'(\ )$ are shown as functions of $(V_0/v_i)\sin\zeta$ in Fig. 30, and in Fig. 31 we give the angular dependences of the components E_r and E_ζ of the field and of the total field $E = (E_r^2 + E_\zeta^2)^{1/2}$ constructed by means of Fig. 30 for $V_0/v_i = 8$. In Fig. 31 next to the extremal points we have written down the values of of the functions $B_0(\ )$ and $B_{0\zeta}'(\ )$.[37] It can be seen from Fig. 31 that in the far zone of a large body the angular distribution of the electric field is fairly complicated, having a number of petals.

The far zone of the wake of a small body ($\rho_0 \ll D$), or rather, a point charge, has been studied theoretically in some detail (Kraus, Watson;[44] Pitaevskii, Kresin;[45] Bud'ko;[37,46] and Vas'kov[47]). At great distances from the body, not only the form of the body but also to a certain extent its size cease to play an important role. Therefore, the results of these investigations are of great interest, since they can be used for comparison with the results of different experiments. The theory of the wake of a small body is also interesting because the corresponding calculations can be made in this case with allowance for the charge of the body, which however it is true must be weakly charged, provided the above conditions (2.13) are satisfied.

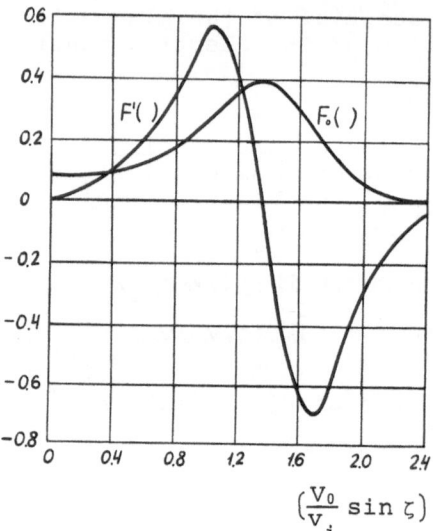

Fig. 30

In the absence of a magnetic field, the perturbation of the electron density in the far zone of a small body is described by the formula

$$\delta N = (V_0^2 B^2 / 4\pi v_i^2) \ln(1/B)$$
$$\times (D^2/r^2) [(2\pi r_B/r) B_1 (\frac{V_0}{v_i}, \zeta) + B_2 (\frac{V_0}{v_i}, \zeta)] \qquad (2.30)$$

where

$$B = (e|\phi_0|\rho_0 v_i)/\kappa TDV_0 \ll 1$$

(see (2.13) above),

$$r_B = 4(V_0/v_i)D/B \ln(1/B), \qquad (2.31)$$

and the nature of the variation of the angular functions $B_1(\ )$ and $B_2(\ )$ at different distances from the body is shown in Fig. 32 (Bud'ko[37]). It follows from (2.29) that there are two characteristic zones of the wake of the body: for distances $r \ll r_B$ and $r \gg r_B$. In the first zone, i.e., at distances appreciably shorter than the characteristic distance r_B, the perturbation δN increases in proportion to $1/r^3$, while in the far zone it decreases

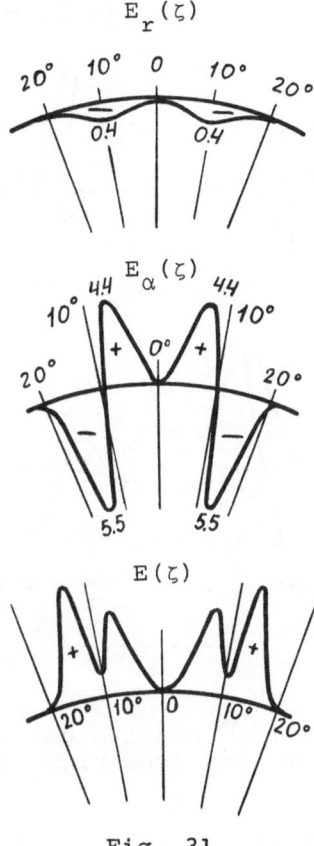

Fig. 31

as $1/r^2$. With changing distance, the angular structure
of the perturbation is gradually deformed. We have already noted above that the basic features of the theoretical dependences δN of a point charge were observed
in the experiments of ref. 40, described above, at rel-
atively short distances from a weakly charged large
body (see Figs. 21 and 22).

In the zone of the distances $r \gg r_B$, where δN increases in proportion to $1/r^3$, the perturbation of the electron density in a nonisothermal plasma can be expressed
in a simple manner analytically in terms of the universal

$$\rho_0 \ll D, \quad H_0 = 0, \quad \phi_0 < 0, \quad V_0/v_i = 8.$$

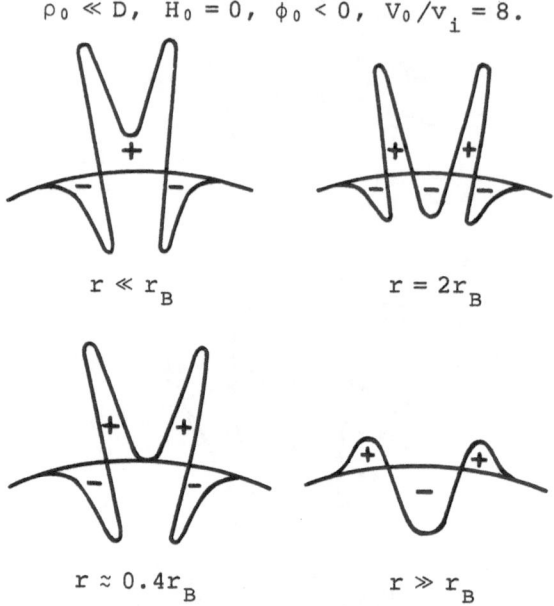

$r \ll r_B$ $r = 2r_B$

$r \approx 0.4r_B$ $r \gg r_B$

Fig. 32

function $F_0\left((V_0/v_i)\sin\zeta\right)$ and its derivative, which describe δN for a large body (see Figs. 26 and 29), namely (Bud'ko[37])

$$\delta N(r,\zeta) = -(V_0^2/v_i^2\pi r^3)[T_e/(T_e + T_i)] \qquad (2.32)$$
$$\times [(1 - 3\cos^2\zeta)F_0 - (V_0/v_i)\sin\zeta\cos^2\zeta\, F_0'].$$

In a magnetized plasma, the perturbation of the electron density averaged over the direction normal to the plane (V_0, H_0) is equal to (Vas'kov[47])

$$\delta N(r,\zeta,\theta_0) = (\pi V_0/v_i^2)B^2\ln(1/B)(D^2/\rho_{Hi}r) \qquad (2.33)$$
$$\times [(r_B/r)B_{1H}\left(\frac{V_0}{v_i},\zeta,\theta_0\right) + B_{2H}\left(\frac{V_0}{v_i},\zeta,\theta_0\right)].$$

The structure of formula (2.33) is similar to that of formula (2.31). However, in this case the perturbation δN at distances $r \ll r_B$ decreases as $1/r^2$, and as $1/r$ for $r \gg r_B$. The corresponding angular dependences of

δN at different distances from the body are shown in
Fig. 33.   They have the same features, although less
well expressed, as the angular functions in the case
when $H_0 = 0$.

§10. Perturbations of the Plasma in the Neighborhood of
Bodies in a State of Quasirest ($V_0 \lesssim v_i$)

As we have seen above (see Tables 1 and 2 of §2)
conditions are realized in the near-Earth plasma such
that satellites or rockets move in some of the regions
of this plasma with velocities V_0 that are not only
comparable with the thermal velocities v_i of the ions
but are even less than v_i.   Under these conditions, the
physical processes in the neighborhood of such a body
are radically different in a number of respects from
those described above for the case $V_0 \gg v_i$.   When $V_0 <
v_i$, the decisive factor affecting the processes in the
plasma is not the velocity of the body but rather its

$$\rho_0 \ll D, \quad H_0 \neq 0, \quad \zeta = \pi/2, \quad \phi_0 < 0, \quad V_0/v_i = 8$$

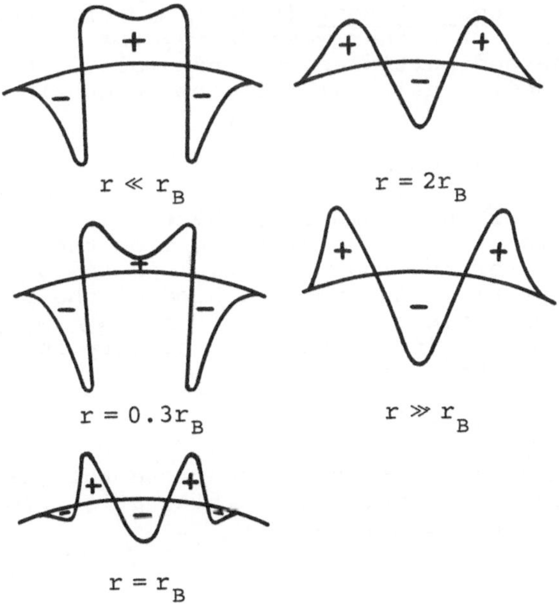

Fig. 33

potential, the reflective properties of its surface,
and, of course, also the magnetic field and linear dimensions ρ_0 of the body. In the transition cases, namely,
when $V_0 \sim v_i$ or $V_0 > v_i$, some of the phenomena are qualitatively similar to those that occur when $V_0 \gg v_i$, but
they can of course differ from them quantitatively by a
considerable amount. Questions of plasma diagnostics
by means of probe measurements of different kinds in
the case of slowly moving bodies take on a specific
nature and require a special treatment. This problem is
an independent problem and a major part of the physics
of probe measurements in the plasma, and its study would
go beyond the scope of this review and take up too much
space. In this section it is however advisable to consider briefly some of the results of theoretical investigations.

When $V_0 \lesssim v_i$, as in the case when $V_0 \gg v_i$, the most
complete results of calculations are available for the
case when the linear dimensions of the body satisfy ρ_0
$\gg D$ or $\rho_0 \ll D$. Since we are here concerned with low
velocities of the body, the basic processes in these
cases are those in the neighborhood of a body at rest.

1. Small Body at Rest ($\rho_0 \ll D$, $V_0 = 0$). For velocities $V_0 \ll v_i$, or rather when $V_0 = 0$, the nature of the
paths of charged particles, and naturally, the distribution of their density $N(r)$ in the neighborhood of the
body depends on the total energy:

$$\varepsilon(r) = Mv^2/2 + e\phi(r) \qquad\qquad (2.34)$$

and the sign of the potential, where v is the velocity
of the particle, r is the distance from the arbitrarily
chosen center of the body, and $\phi(r)$ is the distribution
of the potential in the neighborhood of the body. Two
species of particle can arise: f i n i t e, $N_{fin}(r)$, and
i n f i n i t e, $N_{inf}(r)$. Thus, the density in the neighborhood of the body is

$$N(r) = N_{fin}(r) + N_{inf}(r). \qquad\qquad (2.35)$$

The f i n i t e particles are those that have closed
orbits around the body. Such orbits arise, for example,
for the Coulomb field if

$$\varepsilon(r) < 0 \qquad\qquad (2.36)$$

in the case of an attractive potential: positive for

electrons and negative for ions. Since, as we have already mentioned above in §6, bodies almost always acquire a negative potential in the plasma, the finite particles are here basically ions. However, finite orbits
can arise only under the influence of collisions between
particles -- for the capture of particles that spiral
round the body it is necessary that they should lose
some of their energy. In the absence of collisions,
the attracted particles impinge on the body and are
absorbed by its surface and are not reflected from it.
It is readily understood that the density of finite particles may increase appreciably about a body because
the particles gradually accumulate in its neighborhood.
In the case of an equilibrium distribution,

$$N_{fin} = N_0 \exp\left(\left|e\phi(r)\right|/\kappa T\right),\tag{2.37}$$

and $N_{fin} \gg N_0$ for $\left|e\phi\right| \gg \kappa T$. For electrons attracted
to a Coulomb center, when they collide with neutral
particles, we have at distances satisfying $r > \sqrt{M/m}\rho_0$

$$N_{fin} = (4/3\sqrt{\pi})N_0 \left(\left|e\phi(r)\right|/\kappa T\right)^{2/3}\left(2\left|e\phi(r)\right|/5\kappa T + 1\right)\tag{2.38}$$

(Gurevich[48]), where $\phi(r) = Q_0/r$ (Q_0 is the charge of the
Coulomb center). It can be seen from (2.38) that when
$\left|e\phi\right| \sim 2\kappa T$ the density of the finite particles already
increases appreciably, $N_{fin}/N_0 \sim 4$. The calculations of
the density of finite particles for the majority of cases
that are of interest under the conditions with which we
are concerned are very complicated and have not been
completed. In particular, this is due to the need to
allow for collisions. We shall therefore restrict ourselves to the brief comments made here concerning these
particles. It must also be borne in mind that in the
majority of cases in the media in which we are interested, especially at sufficiently great distances from the
Earth, where the condition $V_0 \ll v_i$ is realized, collisions between particles generally play a small role because the time taken for particles to accumulate on finite orbits is very long.

In the case when the energy of particles at a given
point around the body, for example, for a Coulomb field,
satisfies

$$\varepsilon(r) > 0,\tag{2.39}$$

the particles have open, i n f i n i t e paths. For infinite particles, $N_{inf}(r)$, to arise the field may be
either a t t r a c t i v e o r r e p u l s i v e. Thus, for
a negative potential $\phi_0 < 0$ of the body the field is attractive for ions. We denote the corresponding infinite
particles by N_{inf}^+. When $\phi_0 < 0$, the field is repulsive
for electrons, and we shall denote the corresponding
particles by N_{inf}^-.

For small bodies at sufficiently small distances
from them $(r \lesssim D)$ simple analytic formulas have been obtained for N^+ (Gurevich;[49] Al'pert, Gurevich, Pitaevskii[5]), namely,

$$N_{inf}^+/N_0 = (x/\pi)[1 + (1 - \rho_0/r)^{\frac{1}{2}}] + \frac{1}{2}[1 - \Phi(\sqrt{x})]\exp(x)$$
$$+ \frac{1}{2}[1 - (\rho_0/r)^2]^{\frac{1}{2}}[1 - \Phi(\sqrt{xr/(\rho_0 + r)})]\exp[xr/(\rho_0 + r)],$$
$$\tag{2.40}$$

where

$$x = e\phi_0\,\rho_0/\kappa Tr, \quad \phi \ll \kappa TD/e\rho, \quad D/\rho_0 \gg 1, \tag{2.41}$$

and

$$\Phi(a) = (2/\sqrt{\pi})\int_0^a \exp(-u^2)\,du \tag{2.42}$$

is the error function. For a Coulomb center $(\rho_0 \to 0)$
whose charge is equal to Q_0, when $\phi(r) = Q_0/r = \phi_0\,\rho_0/r$
exactly,

$$N_{inf}^+/N_0 = 2\sqrt{x/\pi} + \exp(x)[1 - \Phi(\sqrt{x})]. \tag{2.43}$$

In the regions of the plasma where the field strength
is low $(e\phi(r)/\kappa T \ll 1)$ though the plasma is not strongly
perturbed, the density of attracted particles N_{inf}^+ is
described by the even simpler formula

$$N_{inf}^+/N_0 = 1 + |e\phi(r)|/\kappa T - (\rho_0/r)^2(|e\phi(r)|/\kappa T + \tfrac{1}{2}). \tag{2.44}$$

Formula (2.44) overlaps well with formula (2.40), which
holds in the region where $|e\phi(r)|/\kappa T \lesssim 1$ but the field is
still a Coulomb one (Gurevich[48], [49]). The dependences
N_{inf}^+/N_0 in the whole of the perturbed region are shown
in Fig. 34 for $D \approx 14\rho_0$ and different values of the potential ϕ_0 of the body. It can be seen that near the
body the density N_{inf}^+ of the attracted particles increas-

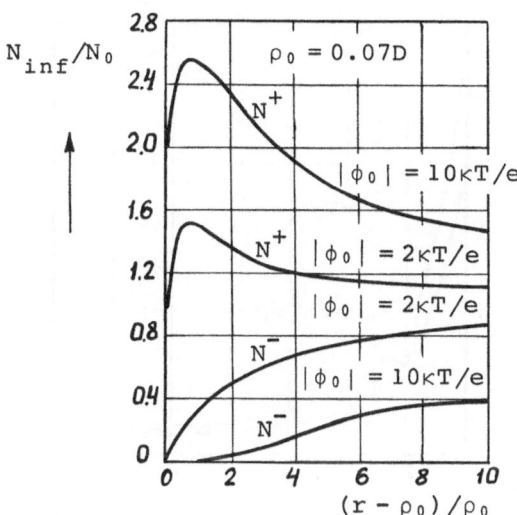

Fig. 34

es considerably. For a Coulomb center, the values of N^+_{inf} are increased even more. For example, when $|e\phi_0|/\kappa T \gg 1$ it follows from (2.43) that $N^+_{inf}/N_0 \approx (2/\sqrt{\pi})\sqrt{|e\phi_0|/\kappa T}$ and when $|e\phi_0|/\kappa T = 10$ that $N_{inf}/N_0 = 4$.

The dependence $\phi(r)$ outside the Coulomb zone is described by fairly complicated formulas that we shall not give here (see ref. 5). In the region of Debye shielding $(r < D)$ it is obvious that $\phi \sim 1/r$. When $D < r \lesssim D \ln (D/\rho_0)$, the potential $\phi(r)$ decreases exponentially and at large distances, when $r > D \ln (D/\rho_0)$, the potential behaves as $\phi(r) \sim 1/r^2$. For $D \approx 14\rho_0$, numerical results of calculations of $\phi(r)$ for the values $|e\phi_0|/\kappa T = 2$ and 3 are given in Fig. 35.

The density of infinite repulsed particles is described not by (2.40) and (2.43), respectively, but by the formulas

$$N^-_{inf}/N_0 = \tfrac{1}{2}\{1 + \Phi(\sqrt{(r-\rho_0)x/\rho_0}) + [1 - (\rho_0/r)^2]^{\tfrac{1}{2}} $$
$$\times [1 - \Phi(r\sqrt{x/\rho_0(\rho_0+r)})] \exp[x\rho_0/(\rho_0+r)]\} \exp(-x),$$

$$(2.45)$$

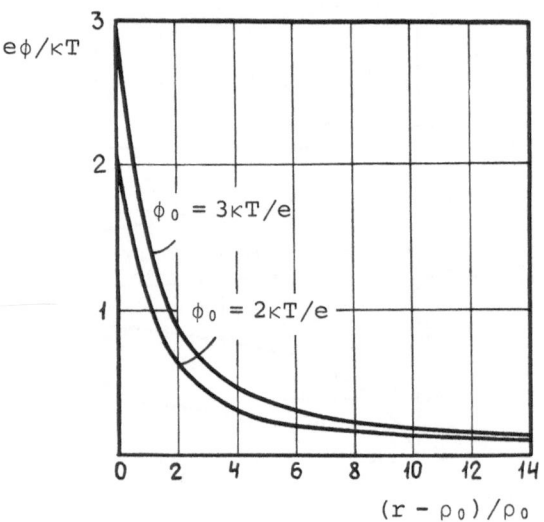

Fig. 35

$$N_{inf}^-/N_0 = \exp(-x).\tag{2.46}$$

In Fig. 34 we have plotted the curves of N_{inf}^-/N_0 for two values of the potential of the body. Near the surface of the body, $N_{inf}^- \ll N_0$, and right up to distances $(1-2)\rho_0$ the density of the infinite repulsed particles, N_{inf}^-, is half or less than half the unperturbed density N_0 of particles.

2. Large Body at Rest ($\rho_0 \gg D$, $V_0 = 0$). An important ant property of the structure of the plasma in the neighborhood of a large body is the formation in a certain region around its surface of a s h i e l d i n g b o u n d a r y l a y e r, because of the strong influence of Debye shielding. In this layer, the condition of quasineutrality of the plasma is strongly violated. At the same time, for not too high potentials of the body, namely if

$$|\phi_0| \lesssim (\kappa T/e)(\rho_0/D)^{4/3},\tag{2.47}$$

the boundary of the shielding layer occurs at $r - \rho_0 \sim D$ (r is measured, as everywhere above, from the center of the body). In the opposite case

$$|\phi_0| \gg (\kappa T/e)(\rho_0/D)^{4/3}, \qquad\qquad (2.48)$$

the boundary of the shielding layer is at $r - \rho_0 \sim \rho_0$.
The criteria (2.47) and (2.48) are thus a measure of a
l o w or a h i g h p o t e n t i a l of the body and determine
the boundary between the near and the far zone of a
large body that is at rest or is moving slowly. The
nature of the variation of the particle density changes
strongly across this boundary. The formula by which it
is described is the same, but for very different laws
of variation of $\phi(r)$. Thus, the densities of infinite
attracted and repulsed particles for $\rho_0 \gg D$ are, respectively, (Gurevich[49,5])

$$N^+_{inf}/N_0 = \tfrac{1}{2}\{\exp(y)[1 - \Phi(\sqrt{y}) + (2/\sqrt{\pi})\sqrt{y}]$$
$$\qquad\qquad\qquad (2.49)$$
$$+ [1 - (\rho_0/r)^2]^{\tfrac{1}{2}} \exp[(y\rho_0^2/r^2 - y_1)r^2/(\rho_0^2 - r^2)]\}.$$

$$N^-_{inf}/N_0 = \tfrac{1}{2}\exp(y)\{1 + \Phi(\sqrt{y_0 - y}) + \sqrt{1 - \rho_0^2/r^2}$$
$$\qquad \times [1 - \Phi(A^{\tfrac{1}{2}})\exp A]\}, \qquad\qquad (2.50)$$
$$A = (\rho_0^2 y_0 - r^2 y)/(r^2 - \rho_0^2),$$

where

$$y_0 = |e\phi_0|/\kappa T,$$
$$y_1 = |e\phi_1(r)|/\kappa T, \qquad\qquad (2.51)$$
$$y = |e\phi(r)|/\kappa T$$

and $\phi_1(r)$ is the potential of the field on the boundary
of the shielding layer.

It is only possible to calculate the potential $\phi(r)$
by the methods of numerical integration. Corresponding
results of calculations for $|e\phi_0|/\kappa T = 10$ and $\rho_0 \gg D$ for
the potential $\phi(r)$ in the near and far zone of the body
and the dependences of the densities of attracted and
repulsed infinite particles obtained from these data are
shown in Figs. 36 and 37. Naturally, the curves shown
in Fig. 36 fit fairly well onto the start of the curves
shown in Fig. 37, i.e., at the boundary of the shielding
layer, which in Fig. 37 has been taken to correspond to
the coordinate origin since $D \ll \rho_0$. An important physical feature of the structure of the plasma in the neigh-

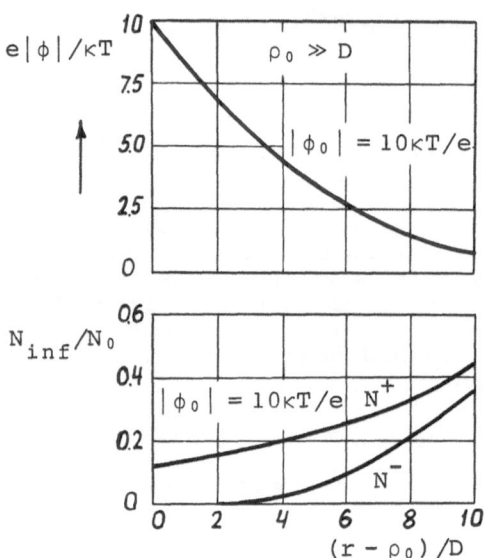

Fig. 36

borhood of a large body is the decrease in the density
of not only the repulsed infinite particles. This is
explained by the fact that the velocity of the attracted particles is considerably increased in the shielding layer, and their flux is conserved. However, the
density of infinite particles can also be increased
around the body if its surface has a good reflective
capacity. In the case of absolute reflection of particles, we can see from Fig. 38, in which we have plotted
the results of corresponding numerical calculations for
$|e\phi_0|/\kappa T = 5$ and 10, that $N^+_{inf} > (10-20)N_0$. Such conditions are seldom realized in experiments on space probes.
However, intermediate cases, or local effects of almost
complete reflection of particles, which complicate the
effects in the neighborhood of bodies of complicated
structure, such as artificial satellites or space probes,
can be observed in individual cases. These circumstances
must be borne in mind when one is planning experiments
or analyzing the results of other experiments.

     3. Slowly Moving Bodies ($V_0 \sim v_i$, $V_0 < v_i$, or $V_0 > v_i$;

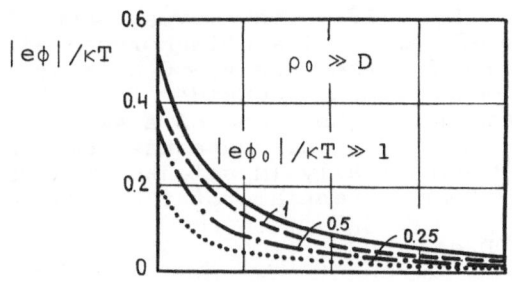

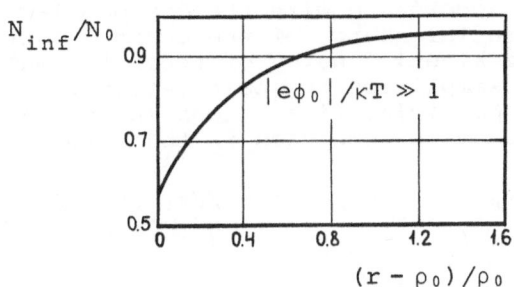

Fig. 37

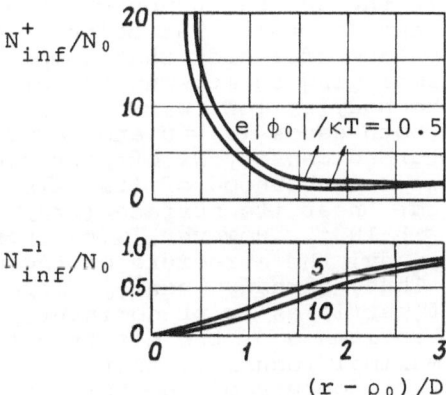

Fig. 38

$\rho_0 \ll D$ and $\rho_0 \gg D$). If a body moves, even at low velocities, the effects in its neighborhood differ radically from those for the case $V_0 = 0$ even at low potentials of the body. Let us consider first the corresponding effects in the neighborhood of a small spherical body of radius $\rho_0 \ll D$; this case has been fairly fully investigated theoretically (Mosalenko,[50] Knyazyuk, Mosalenko[51]). For low potentials of the body,

$$|\phi_0| \ll \kappa TD/e\rho_0, \qquad (2.52)$$

the electric field in the plasma is weak, and it can be assumed that in the whole range of distances it also decreases in accordance with the Coulomb law, since when $r \gtrsim D$ the potential energy of the charged particles is less than the kinetic energy of their thermal motion. This greatly simplifies the calculations. The density of attracted particles (i.e., ions in the case $\phi_0 < 0$) can be expressed in this case in the form

$$N(r,\zeta)/N_0 = N^+_{inf}(r)/N_0 + (V_0/\sqrt{\pi}v_i)f(V_0/v_i,r,\zeta,\phi_0),$$
$$(2.53)$$

where $N^+_{inf}(r)$ is the density for $V_0 = 0$ (corresponding formulas are given in the foregoing subsection), and the function $f(\ )$ can be expressed in quadratures and calculated by numerical methods. The results of calculations for the case $V_0 \sim v_i$ and different values of ϕ_0 (Knyazyuk, Mosalenko[51]) are shown in Fig. 39. In Fig. 39 the continuous curves are those of equal values of $N(r,\zeta)/N_0$, while the dashed curves are those for $N^+_{inf}(r)/N_0$ for a body at rest ($V_0 = 0$). The calculations were made for the surface of a body that completely neutralizes the ions impinging on it, and therefore, when the potential of the body is low ($\phi_0 \sim 10^{-2}\kappa T/e$), the density of the ions in both cases ($V_0 = 0$ and $V_0 \sim v_i$) is less than the unperturbed density N_0 of particles. Behind the body, in the neighborhood of its axis, a depletion region is formed; near its surface ($r/\rho_0 \approx 1$, $\zeta = 0$) we have $N(1.0)/N_0 \approx 8 \cdot 10^{-2}$. However, with increasing potential of the body, the structure of the plasma changes considerably. Thus, already when $\phi_0 \approx \kappa T/e$ (see Fig. 39) accumulation of particles is predominant, $N(r,\zeta) > N_0$, because of the influence of the electric field. At the same time the maximal focusing, that is, the maximal value of $N_{max}(r,\zeta)$, occurs on the axis behind the body at a distance z_{max} from the surface of the body, this distance depending on the potential ϕ_0. When $\phi_0 \approx \kappa T/e$,

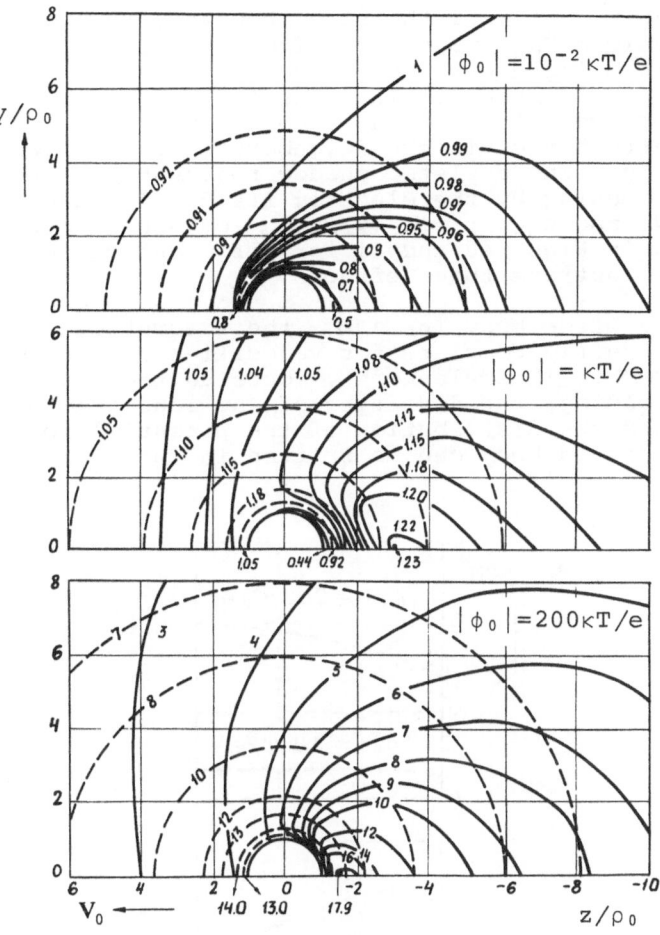

Fig. 39

we have $N_{max}/N_0 \approx 1.23$ and $z_{max}/\rho_0 = 3$. However, in the
immediate proximity of the surface of the body for small
values of ϕ_0 a small depletion region is also formed.
Thus, when $\phi_0 \approx \kappa T/e$ at a distance $z/\rho_0 = 1$ we have $N(1.0)/N_0$
≈ 0.21. However, with increasing potential of the
body, the depletion region gradually disappears, and
the extent to which the particles accumulate increases,
and the maximum of N approaches the body. When $\phi_0 =$
$200\kappa T/e$, we have $N_{max}/N_0 \approx 1.79$ and $z_{max}/\rho_0 \approx 1.3$. With

increasing velocity of the body, however, the position
of the maximum N_{max} moves away from the surface of the
body, this being due to the increase in the depletion
region -- the "shadow". Thus, as in the case $V_0 \gg v_i$,
two effects compete, and these determine the structure
of the perturbation in the neighborhood of a slowly
moving body: the depleted shadow region behind the body
and the focusing by the electric field. The dependences
of N_{max}/N_0 and z_{max}/ρ_0 on $(V_0/v_i)^2$, obtained in ref. 51
and shown in Figs. 40 and 41 for different values of
$e\phi_0/\kappa T$, illustrate these effects.

For a large body ($\rho_0 \gg D$), the theoretical calculations are fairly complete for velocities of the body satisfying $V_0 \ll v_i$ (Moskalenko[52]). Outside the region of
Debye shielding, the densities of the ions and the electrons and the distribution of the potential for a negatively charged body can be written in this case in the

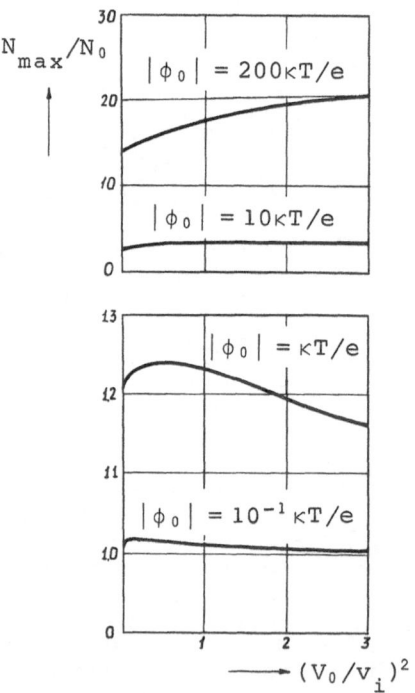

Fig. 40

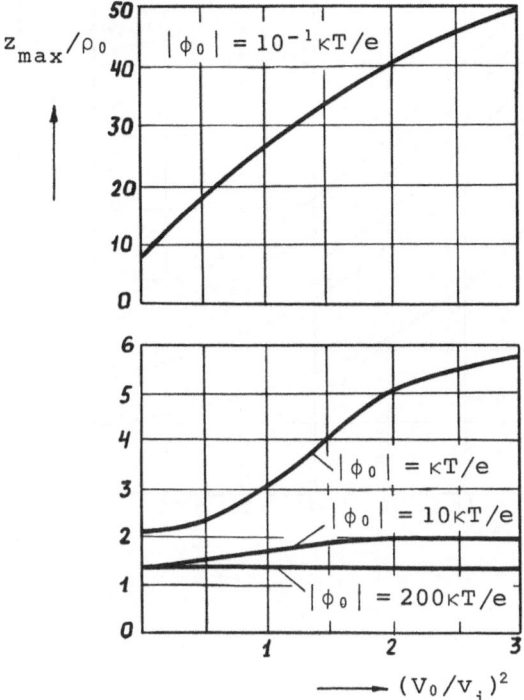

Fig. 41

form

$$N_i(r,\zeta)/N_0 = N_i(r)/N_0 + (V_0/v_i)\,f_N\!\left(\frac{r}{\rho_0},\frac{e\rho_0}{\kappa T}\right)\cos\zeta,$$

$$N_e(r,\zeta)/N_0 = N_e(r)/N_0 + (V_0/v_i)\,f_N\!\left(\frac{r}{\rho_0},\frac{e\phi_0}{\kappa T}\right)\cos\zeta,$$
$$\tag{2.54}$$
$$\phi(r,\zeta) = \phi(r) + (V_0\,\kappa T/v_i e)\,f_\phi\!\left(\frac{r}{\rho_0},\frac{e\rho_0}{\kappa T}\right)\cos\zeta,$$

where $N_i(r)$, $N_e(r)$, and $\phi(r)$ are the values of these quantities for $V_0 = 0$; formulas for them are given in the foregoing subsection for infinite attracted (ions) and repulsed (electrons) particles. Graphs of the functions $f_n(\ )$ and $f_\phi(\ )$, which are calculated in ref. 52 by numerical methods, are given for different values of $e\phi_0/\kappa T$ in Figs. 42 and 43.

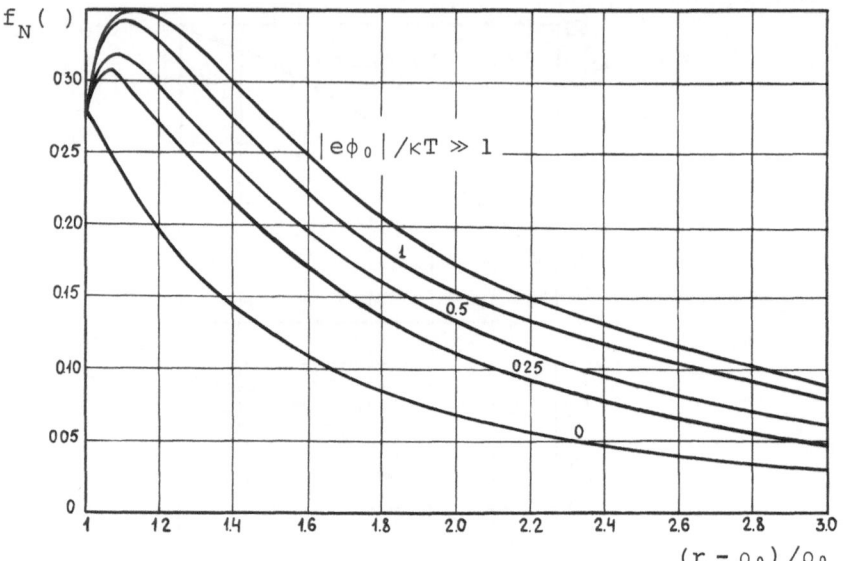

Fig. 42

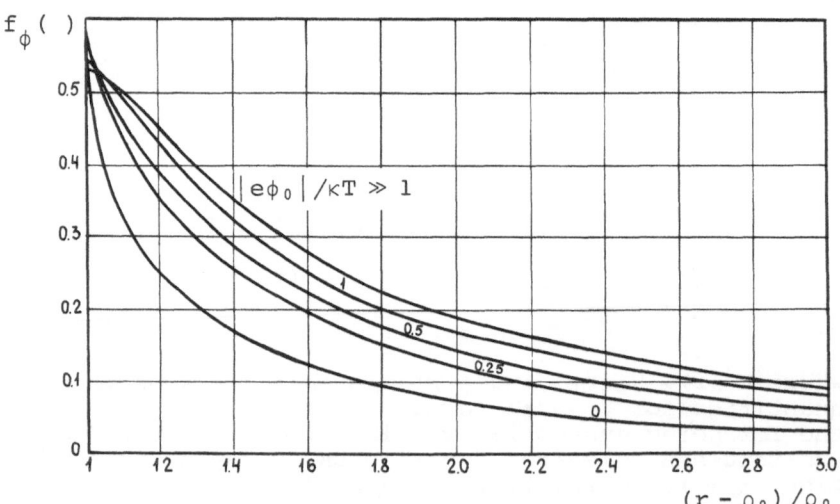

Fig. 43

§11. Scattering of Radio Waves on the Wake of a Rapidly
Moving Body

As we have seen in §9, the wake of a rapidly moving body is an inhomogeneous cloud extended behind the
body and "carried along" by the body. This cloud is
described by the perturbation $\delta N(r)$ of the electron density and, therefore, by the perturbation of the plasma
permittivity surrounding the body. It is easily seen
that the wake of the body is extended to a length of the
order of the mean free path of the particles -- at these
distances, $\delta N(r)$ is gradually smeared out. The mean free
path, which increases gradually in the ionosphere with
altitude, attains, because of the reduced number of collisions between particles, hundreds of meters and sometimes kilometers or more. Therefore, although $\delta N(r)$ is
small in the far zone of the wake, which makes up the
principal part of its length, we must expect that under
certain conditions there will be appreciable scattering
of electromagnetic waves on this long inhomogeneous cloud
formed behind the body. The first simple estimates of
this effect (Al'pert[53]) showed that if one takes into
account the influence of the external magnetic field of
the Earth, which leads to a cylindrical structure of
the wake, the effective differential cross section $d\sigma$
for the scattering of radio waves on the wake may at-
tain 100 m^2 or more and exceed the cross section of scattering of the body itself by factors of tens or hundreds.
Subsequent rigorous and detailed theoretical investigations of this effect (Al'pert, Pitaevskii[54, 5]) did indeed
confirm that under certain conditions the scattering
cross section of the wake of a body may have large values and that if the body moves in the ionosphere, below
its principal maximum near the caustic, $d\sigma$ may even be
or order 10^4 m^2 or more (Gurevich, Pitaevskii;[55] Vas'-
kov[56]). At the same time, to the best of our knowledge,
there have not yet been any reports in the literature
of experimental results that convincingly confirm this
effect. In particular, as we shall see below, this is
because $d\sigma$ has a petal structure. The width of the
part of its principal petal, which corresponds to large
values of $d\sigma$, is only of the order of fractions of a
degree. Therefore, because of the rapid motion of the
body, the scattered field may be detected on the Earth's
surface only in the form of short-period bursts of duration ~1 sec or less. Thus, for the experimental investigation of this effect one requires specially set
up and sufficiently accurate experiments, and one cannot be confident that the various reports in the lit-

erature concerning observations of large values of the
scattering field of radio waves from artificial satellites (see, for example, refs. 57 and 58) are indeed the
results of the experimental observation of this phenomenon. However, the effect of scattering of radio
waves from the wakes of bodies is of undoubted interest not only from the point of view of studying the physics of the interactions between moving bodies and the
plasma but also from the point of view of practical
applications in a number of respects. The results of
theoretical investigations of this effect are set forth
briefly below.

To calculate the scattering on the wake of a body,
one can use perturbation theory, since the values of
$\delta N(r)$ in the part of the wake that plays a role in the
scattering of radio waves are small compared with the
electron density N_0 of the unperturbed plasma. Following the method of calculations given in ref. 59 (Landau,
Lifshits), one finds that the effective differential
cross section of scattering $d\sigma$ on the wake, which determines the fraction of energy of the electric scattering
field E_s in unit solid angle do is equal to

$$d\sigma = (|E_s|/|E_0|^2) r^2 do$$
$$= (1/16\pi^2)(\omega_0/c)^4 |\delta Nq|^2 N_0^{-2} \sin^2 \psi do .$$ (2.55)

In formula (2.55), ψ is the angle between the vector of
the electric field E_0 of the incident wave and the wave
vector k_s of the scattered wave; $\omega_0^2 = 4\pi N_0 e^2/m$ is the
electron plasma frequency; and δNq is the Fourier trans-
form of the spatial distribution of the perturbation
of the electron density $\delta N(r)$. Thus, the differential
scattering cross section $d\sigma$ is expressed in terms of
the perturbation of the electron density in the q space,
namely, in terms of the quantity

$$\delta Nq = \int \delta N(r) \exp(-iq \cdot r) dr^3 .$$ (2.56)

At the same time,

$$q = k_s - k_0 ,$$ (2.57)

where k_s and k_0 are, respectively, the wave vectors of
the scattered and the incident wave. The solution of
the stationary ($\partial f/\partial t = 0$) kinetic equations (1.11) in
the quasineutral approximation with allowance for (1.16)
enables one to obtain the perturbation δNq in the form

of closed formulas (Pitaevskii[60,5]). This facilitates exact calculations of $d\sigma$, which, however, can be carried through to the end only by the method of numerical integration.

The upshot is that for a sphere of radius ρ_0 it is found that

$$d\sigma = \frac{\omega_0^4 \rho_0^4 V_0^2}{16c^4 \Omega_{Hi}^2} F(\tau,\xi,\delta,\eta) |G(q\rho_0,\chi)|^2 \sin^2\psi, \qquad (2.58)$$

where χ is the angle between the vectors $\mathbf{q}$ and $\mathbf{V_0}$,

$$\tau = (qV_0/\Omega_{Hi}) \cos\chi$$
$$= (qV_0/\Omega_{Hi})(\cos\theta_0 \sin\theta_q + \sin\theta_0 \cos\theta_q \cos\phi_q),$$
$$\xi = (v_i^2 q^2/4\Omega_{Hi}^2)\sin^2\theta_q, \qquad (2.59)$$
$$\delta = \tfrac{1}{2}(v_i^2 q^2/\Omega_{Hi}^2)\cos^2\theta_q,$$
$$\eta = v_{ii}/\Omega_{Hi},$$

θ_q is the angle between $\mathbf{q}$ and the normal to H_0; ϕ_q is the angle between the planes $(\mathbf{q},\mathbf{V_0})$ and $(\mathbf{V_0},H_0)$; θ_0 is the angle between $\mathbf{V_0}$ and H_0; and v_{ii} is the effective ion-ion collision frequency.

The functions $F(\ )$ and $G(\ )$, which determine the effective differential cross section of scattering $d\sigma$ in (2.58), can only be calculated by the method of numerical integration.[54] With regard to the function $G(\ )$ it is smooth and varies slowly as a function of the angle χ. In addition, it is evident from what follows that $d\sigma$ has maximal values for small values of τ. Therefore, in the calculations one can use the value of $G(\ )$ for $\chi = \pi/2$. In this case

$$G(q\rho_0,\pi/2) = [J_1(q\rho_0)/q\rho_0]^2, \qquad (2.60)$$

where $J_1(\ )$ is a Bessel function.

The basic properties of $d\sigma$ are determined by the function $F(\ )$. We shall therefore call it the s c a t- t e r i n g   f u n c t i o n. It depends on four parameters, which in their turn depend on the velocity $\mathbf{V_0}$ of the body and the angle between this velocity and the magnetic field H_0 of the Earth, on the thermal velocity v_i,

the gyrofrequency Ω_{Hi}, and the ion-ion collision frequency ν_{ii} in the ionosphere. The properties of the
function $F(\ )$ can be seen from the figures and tables
given below, which are calculated for three altitudes
$z = 300$, 400, and 700 km in the ionopshere, where the
effect of scattering on the wake of a body in the ionosphere is evidently most strongly expressed. For a
narrow region in the neighborhood of the maxima of the
differential cross section $d\sigma$ (see below), formula
(2.58) is replaced by the following approximate analytic
formula (Vas'kov[61]):

$$d\sigma = (\omega_0/2c)^4 (\rho_0^4 V_0^2/\nu_{ii}^2) |P_n(\ )|^2 do, \qquad (2.61)$$

where

$$P_n(\ ) = (\sqrt{\pi}/a) W(d) \exp(-\mu) I_n(\mu) \qquad (2.62)$$
$$\times \{2 + i\sqrt{\pi}[n\Omega_{Hi}/a\nu_{ii} + (d + i/a)]W(d)\exp(-\mu)I_n(\mu)\}^{-1},$$

and

$$W(d) = \exp(-d^2)\left\{1 + (2i/\sqrt{\pi})\int_0^t \exp(t^2)dt\right\} \qquad (2.63)$$

is the Kramp function ($W(d) \sim i/d$ for $d \gg 0$). In (2.62)

$$d = B/a + i/a,$$
$$a = |q_{\|}|v_i/\nu_{ii}, \qquad (2.64)$$
$$B = [q \cdot V_0 - n\Omega_{Hi}]/\nu_{ii},$$
$$\mu = \tfrac{1}{2}q_{\perp}^2(v_i^2/\Omega_{Hi}^2),$$

where $n = 1, 2, \ldots$ is the number of the corresponding
maximum of $d\sigma$, $q_{\|}$ and $q_{\perp}$ are the components of q along
and at right angles to the vector H_0; and $I_n(d) > 0$ is
a Bessel function of imaginary argument. Although it is
obvious that formula (2.62) is fairly complicated, it
nevertheless enables one to determine $d\sigma$ without recourse
to methods of numerical integration. However, a complete
analysis of the properties of the scattering cross section was made on the basis of an investigation of the
exact formula (2.58),[54] some numerical results of which
are given below.

The main feature of the scattering function $F(\ )$
is its oscillatory -- multipetal -- nature as a function
of the angle (Fig. 44). If $\theta_0 = 0$, i.e., the body moves
along the magnetic field vector H_0 of the Earth, or in
the case when $\theta_q = 0$, the principal maximum of the scat-

tering function (maximum of zeroth order (0)) corresponds
to the value $\tau = 0$ $(\chi = \pi/2)$, and its lateral maxima (of
order ± 1, ± 2,...) are placed symmetrically with respect
to it. This can be seen in Fig. 44a, in which the
plotted curves correspond to different values of the
parameters ξ and δ. Analysis of the general properties
of the scattering function shows that its principal
maximum has largest values when $\theta_0 = \theta_q = 0$. If $\theta_0 = 0$,
the principal maximum lies in the direction of "specu-
lar reflection" of the wave from the direction of the
magnetic field H_0. This means that the bisector of the
angle (k, k_s), i.e., the vector q, is perpendicular to
H_0. At the same time, since $F(\ )$ does not depend on the
angle θ_q when $\theta_0 = 0$, the function $F(\theta_q)$ forms a surface
of revolution about the vector k_s, along which its
principal "spatial petal" is directed. For $\theta_0 \neq 0$, the
principal maximum corresponds to values $\theta_q = -\theta_0$ (Fig.
44b). The vector q is turned through the same angle
with respect to the normal to H_0. The number of lateral maxima and their values, which decrease rapidly
with increasing number of the petals, depend strongly
on the parameters ξ, δ, and n, which determine the con-
vergence of the corresponding integrals. As is readily
noted (see (2.59)), the function $F(\ )$ depends on the
angle ϕ_q only if $\theta_0 = 0$.

Thus, the scattering effect must be observed when
the body moves along the vector of the magnetic field,
the m a i n   r o l e   then being played by the p r i n c i p a l
m a x i m u m of $F(\ )$. It has fairly large values in ranges
of angles $\Delta\theta_0$ or $\Delta\theta_q$ that amount to only fractions of a
degree. The corresponding dependences of the values of
the principal maximum $F_{0,max}$ (maximum of 0-th order) of
the scattering function on the wavelength λ for $\theta_0 = 0$
and $\theta_q = 0$ are given in Fig. 45, and in Table 3 we have
given the values of $F(\ )$ for different altitudes z and
wavelengths λ in the neighborhood of the principal maximum, and also, for $\lambda = 30$ m and $\theta_0 = 0$, the values of
the first-order maximum, $F_{1,max}$, which, in the considered
case, corresponds to the angle $\theta_q = -5°$.

It can be seen from Table 3 and Fig. 44 that the
field of the scattered wave can be fairly large only
in a small region of angles in which the maximum of zeroth order of the scattering function has large values.
Calculations show that one can adopt the corresponding
angular width of the petal:

$$\alpha \approx 2\Delta\theta_0 \quad \text{or} \quad 2\Delta\theta_q \approx (0.6-0.8)10^{-2} \, \text{rad.} \qquad (2.65)$$

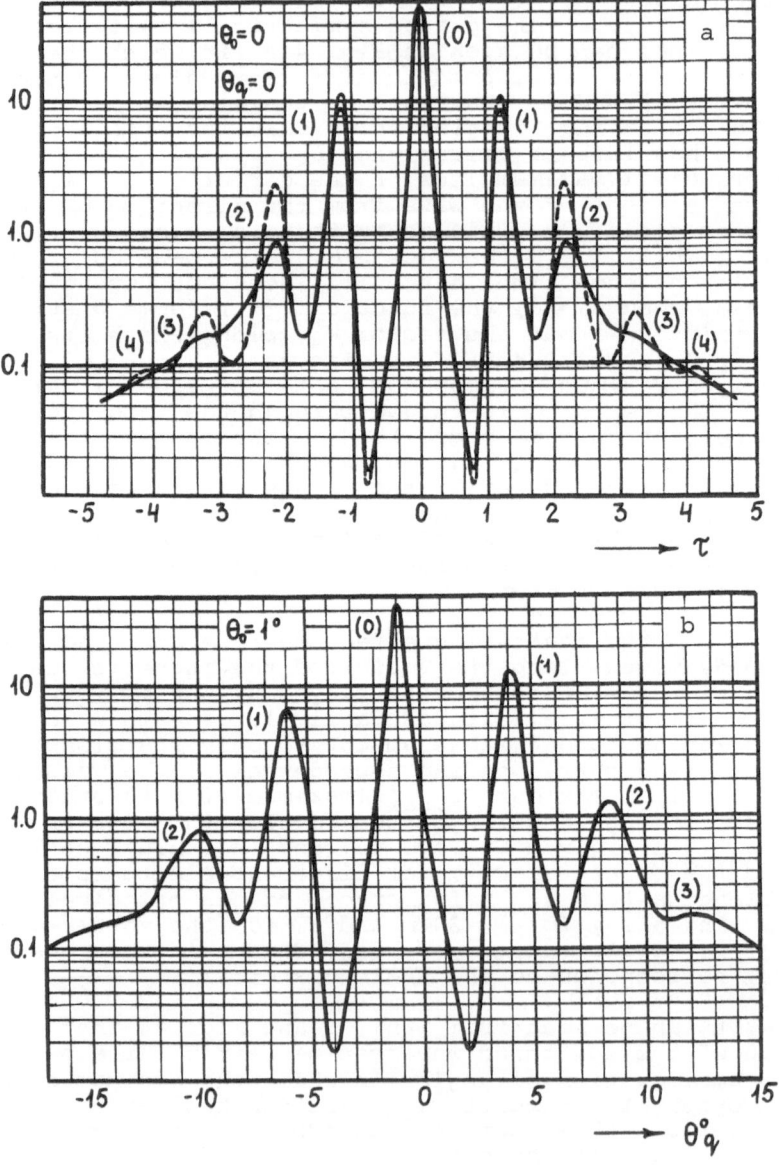

Fig. 44

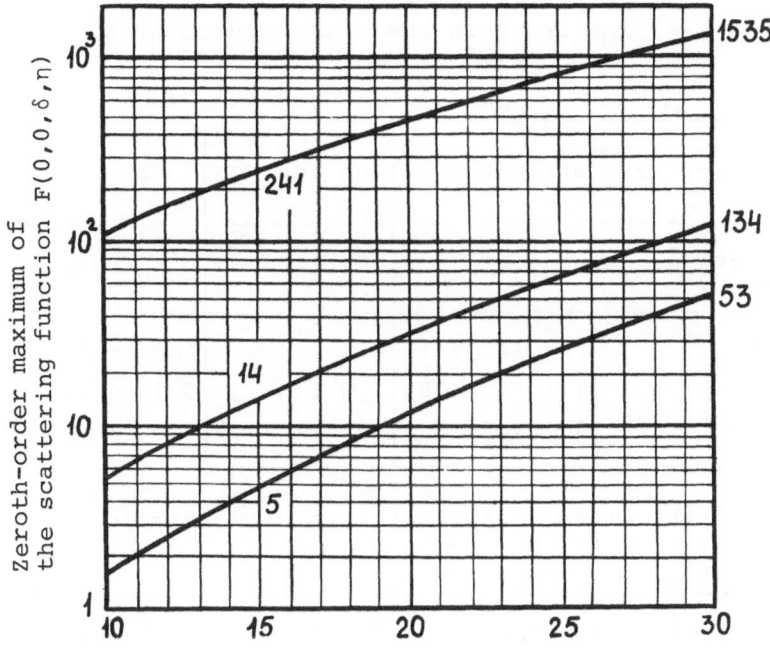

Wavelength λ, m

Fig. 45

Therefore, at a velocity $V_0 \approx 8$ km/sec of the body, the scattered field at the surface of the Earth in the case of motion of the body along H_0 at the above altitudes z "illuminates with a flash" an area on the Earth's surface of size $(\alpha \cdot z)$, continuing to have maximal values only during time intervals of

$$\Delta t \sim [\alpha(\text{rad}) \cdot z]/V_0 \sim 0.4\text{-}1 \text{ sec.} \tag{2.66}$$

The total interval of angles filled with several petals is approximately 15-20°, although the flashes of the ±1st, ±2nd, and higher order are much weaker. At the same time, in the given time intervals Δt the measurable effect of the scattered field in the receiving device is naturally determined not by the maximal values $F_{0,max}$ of the scattering function given in Figs. 45 and Table

TABLE 3. Values of the Zeroth-Order Maximum of the
Function F( ) for $\theta_0 = 0$ in the Neighborhood of the Angle
$\theta_q = 0$

Altitude z, km →	300 400 700	300 400 700	300 400 700
Wavelength λ, km ↓	$\theta_q = 0$	$\theta_q = \pm 0.3$	$\theta_q = \pm 1^\circ$
15	5 14 241	0.5 0.4 -	0.01 0.005 0.001
20	11 31 479	1.8 1.5 0.7	0.1 0.07 0.03
30	53 134 1535	13 11 4.5	1.3 1.0 0.3

Values of the First-Order Maximum of the Function F( )
for $\theta_0 = 0$, $\theta_q = -5^\circ$, and $\lambda = 30$ m.

z, km	300	400	700
F_1, max	10.3	12.6	3.4

3 but by a value obtained by averaging in a certain manner over the given intervals of the angle α. These "average" values depend on a number of factors, in particular,
the properties of the receiving apparatus, the angular
dependence of F( ), and so forth. Approximate estimates
show that one can take

$$\overline{F(\ )} \sim \tfrac{1}{2} F_{1,max}. \tag{2.67}$$

By means of formula (2.58) calculations were made
of the effective differential cross section of scattering, $(d\sigma/do)_{max}$, as a function of the wavelength λ
and the altitudes z (Fig. 46) from the values of $F_{0,max}$
given in Fig. 45. Since $d\sigma \sim \omega_0^4 \sim N_0^2$ (see (2.58), i.e.,
the cross section depends strongly on the electron den-
sity, the effective cross section $d\sigma$ of the wake of a
body is small during the nighttime. Therefore, the cor-

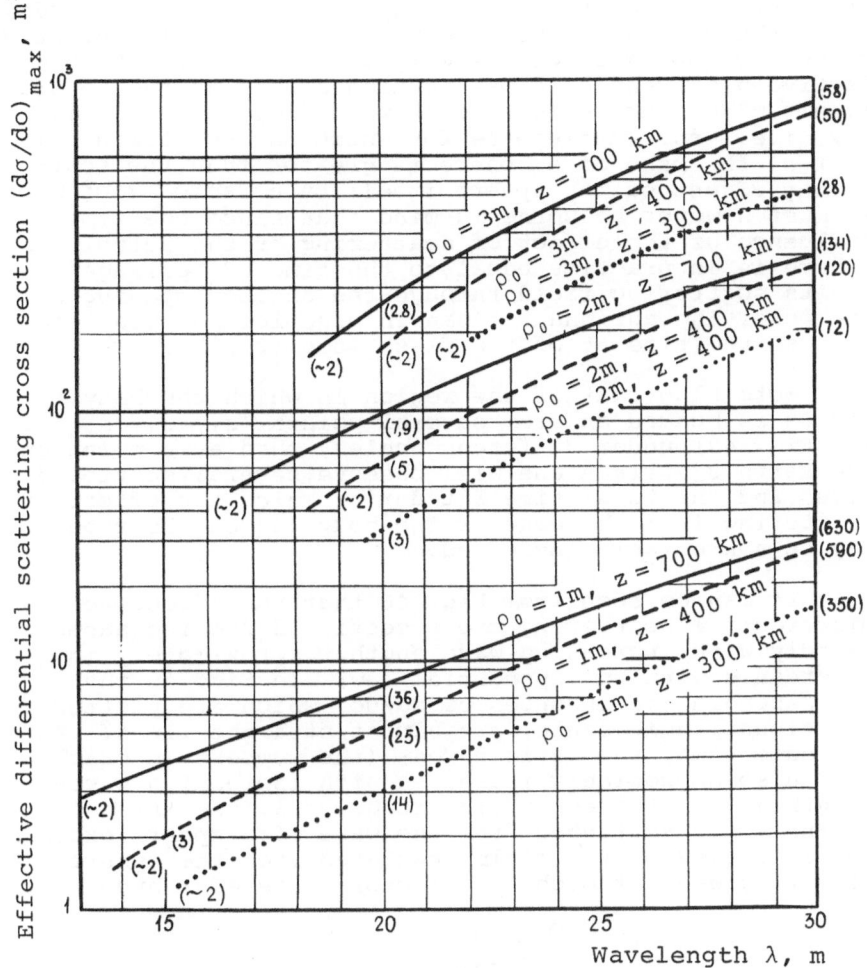

Fig. 46

responding curves in Fig. 46 are given for an average
model of the ionosphere during the daytime. It is also
of interest to compare the effective scattering cross
section of the wake of the body, $d\sigma_0/do$, with the scattering cross section of the body itself. For such comparison, we assumed scattering from a smooth metallic
sphere of radius ρ_0. Other bodies with similar linear
dimensions but a complicated (in particular rough)
structure of the surface have smaller values of $d\sigma_0$. In

Fig. 46, in brackets, we have given the values of $d\sigma/d\sigma_0$,
i.e., the ratios of the effective differential scattering
cross section $d\sigma/do$ of the wake of a sphere to the effective differential cross section $d\sigma_0/do$ of the sphere
itself. The curves in Fig. 46 were plotted for $d\sigma/d\sigma_0$
> 2, i.e., for wavelengths for which the scattering ef-
fect of the wake of a sphere is greater than the scattering effect of the sphere itself by a factor 2 or more.
It must however be borne in mind that since the time
influence of the effect of scattering of the actual
sphere is in practice equal to the time of "passage"
of its scattering field through the angular opening of
the receiving antenna, it can in practice be much long-
er than the time Δt of the flash (see (2.65)).

Note also that if the region in which the body is
moving is beamed on from several points (s_1, s_2, s_3,...;
see Fig. 47) under different angles, then at a point on
the Earth's surface one will observe several scattered
waves and the total time $\Sigma\Delta t$ during which the effect of
scattering from the wake of the body can be observed
may be appreciably increased.

It can be seen from Fig. 46 that the effective
differential scattering cross section $d\sigma/do$ increases
rapidly with increasing wavelength, approximately in
accordance with an exponential law; in Fig. 46 the
corresponding dependences of $d\sigma/do$, which are plotted
on a logarithmic scale, are almost straight lines. At
the same time, the ratio $d\sigma/d\sigma_0$ (the numbers in brackets) also appreciably increases with increasing wavelength λ. In the considered range of λ, the intensity
of the field scattered from the wake of a sphere exceeds
the intensity of the field scattered from the sphere
itself by even as much as a factor of several hundreds.

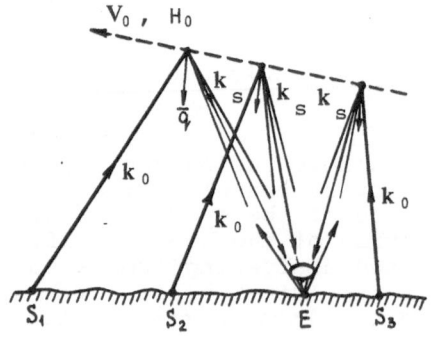

Fig. 47

As can be seen from Fig. 46, the differential cross
section $d\sigma/do$ increases with altitude, although much
more slowly than with increasing λ. The increase of
$d\sigma/do$ with increasing z is explained by the fact that
the decrease in the collision frequency ν_{ii} with altitude is stronger than the influence of the reduction in
the electron density N_0 -- these two have opposite in-
fluences on $d\sigma$.

All the above results of theoretical calculations
were obtained under the assumption that the incident
waves scattered on the wake of the body are plane. If
allowance is made for the incident wave's being spherical, some new features are revealed (Vas'kov[62]) but
these do not significantly modify the properties considered above nor the results of the quantitative calculations of the effective differential cross section in
the case when one can ignore the inhomogeneity of the
ionosphere and the body moves in regions in which the
value of the refractive index of the ionosphere is not
near zero. The influence of the wave's being spherical
reduces in this case to the principal maximum of the
scattering cross section being also cut off as a result
of the sphericity of its front (for a plane wave, the
corresponding divergence of the formulas for $d\sigma$ is cut
off only by the collision frequency ν_{ii}). This is due
to the fact that the effective length of the wake, which
determines the principal part of the intensity of the
scattered field, is restricted to the first Fresnel
zone of radius $\rho_F \sim \sqrt{\lambda s}$, where s is the distance between
the source of the incident waves and the body. However,
sphericity of the wave becomes very important when the
body moves in an inhomogeneous medium and passes through
a region of reflection of the incident wave, where the
refractive index of the region tends to zero, $n^2 \to 0$. In
this case, it has been shown (Gurevich, Pitaevskii[55])
that the values of $d\sigma$ may increase by two orders or more,
especially for small angles of the incident wave. This
effect becomes very strong when the body moves near the
caustic formed by the reflection of the spherical waves
from the inhomogeneous medium. In this case there is
maximal focusing of the waves. To obtain this effect
it is then sufficient to take into account only the inhomogeneity of the ionosphere in altitude, as was done
in refs. 55 and 56.

As is well known, the caustic is the envelope of
the family of rays emitted by the source and it is formed
by the refraction of waves in the inhomogeneous medi-

um. The caustic separates the region illuminated by
waves from the shadow region. It is therefore natural
that the nature of the variation of the scattering cross
section varies on the transition of the body from the
illuminated to the shadow zone and, therefore, depends
on the position of the body relative to the caustic.
Detailed calculations of the effect (Vas'kov[53, 63]) have
led to the following results. The intensity of the
scattered wave is maximal when the source of the incident waves and the observer are at one point, when the
body moves tangentially to the surface of the caustic
and the geomagnetic field is normal to this surface. The
dependence of the effective differential scattering
cross section on the distance r_c for a sphere of radius
$\rho_0 = 1$ m before a caustic, which was calculated in ref.
63 for the region of the ionosphere at the altitude $z =$
250 km ($N_0 \approx 10^6$ cm^{-3}), is shown in Fig. 48. In this
figure, curve a corresponds to a motion of the body that
approaches the caustic on the side of the illuminated
region ($r_c > 0$), curve b to the region when the body,
after insecting the caustic, goes over into the shadow
region ($r_c < 0$), and curve c to $d\sigma/do$ as a function of
r_c when the body moves only in the illuminated region.
If the body passes into the shadow region behind the
caustic, scattering occurs only on part of its wake
that is in the illuminated region. The quantity $z_{eff} =$
V_0 / ν_{ii} in Fig. 48 is the so-called effective length of
the wake of the body, and β is the angle between V_0 and
the normal to the caustic. It can be seen from Fig. 48
that in this specific case the cross section attains
the value 10^4 cm^2 when $r_c = 0$. It should be noted that

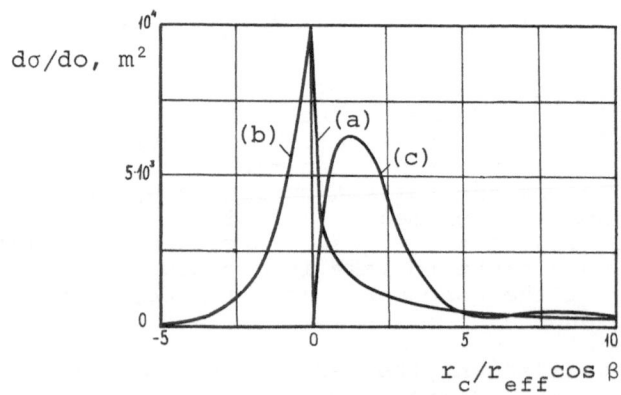

Fig. 48

the dependences of $d\sigma/do$ on $r_c/(z_{eff}|\cos\beta|)$ shown in Fig. 48 were obtained in ref. 63 for the case when $\beta \neq \pi/2$, namely, when the following condition holds:

$$r_{eff}|\cos\beta| > [(\lambda/2\pi)^2 (dn^2/dz)^{-1}]^{\frac{1}{2}}. \qquad (2.68)$$

Since $r_{eff} \sim 2\cdot10^2$ m for $z \sim 250$ km and for $N_0 \sim 10^6$ reflection occurs at a wavelength $\lambda \sim 33$ m, taking dn^2/dz $\sim 10^{-4}$ m^{-1} we find that the curves in Fig. 48 correspond to the case when $|\cos\beta| \gtrsim 0.25$, $\beta \lesssim 76°$. With increasing angle β, as we have pointed out above, the value of $d\sigma/do$ increases; exact calculations for the limit $\beta \rightarrow \pi/2$ are especially complicated and have not been made. For the given value, $r_{eff} \approx 200$ m, and for motion of the body near the caustic strong scattering occurs at distances of order 1-2 km, and therefore the time of the burst of the intense scattering field at the point of observation is $\Delta t \sim 0.1-0.2$ sec. It is interesting that in certain cases when the body moves near the caustic $d\sigma/do$ and, therefore, the scattered field received near the Earth's surface must have an oscillatory nature. Such an effect arises when there is a definite geometrical disposition of the wave front with respect to the caustic, when the sphericity of the wave may cut off -- restrict -- the scattering cross section and there is interference between the waves that come from the region of the caustic and those that are reflected from the Fresnel zone. The corresponding case, calculated for the above conditions in the ionosphere (see ref. 63), is shown in Fig. 49, in which we have plotted $d\sigma/do$ as a function of the distance r_F of the body from the center of the Fresnel zone (ρ_F is, as above, the radius of the first Fresnel zone).

§12 Some Remarks on the Excitation of Waves and the Instability of the Plasma in the Neighborhood of a Rapidly Moving Body.

Very important problems that arise in the study of the flow of plasma around rapidly moving bodies, when $V_0 \gg v_j$, are investigations of:

the conditions of instability of the plasma surrounding the body,

the types of oscillations that are excited in the perturbed region of the plasma and are carried along with the body and,

the possibility of electromagnetic waves being radiated by the perturbed region of the plasma.

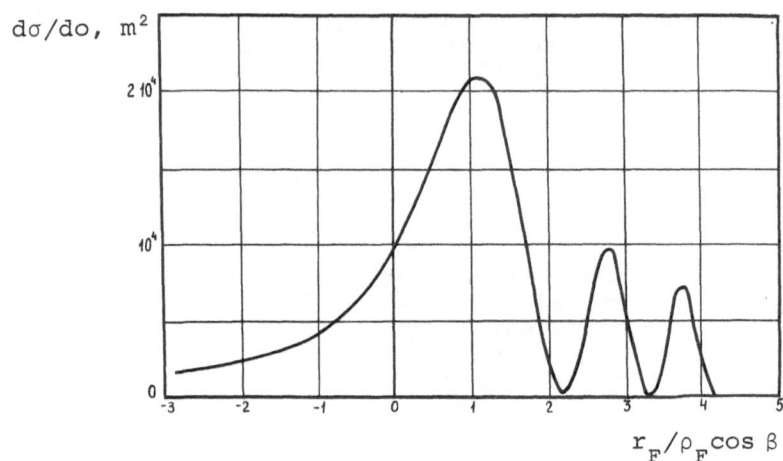

Fig. 49

        Naturally, when formulated in full generality,
these problems require the solution of the kinetic equations (1.11) with allowance for the dependence on the
time t, i.e., including the terms ∂f_e $(\mathbf{r},\mathbf{v},t)/\partial t$ and
∂f_i $(\mathbf{r},\mathbf{v},t)/\partial t$;   of course, it is more convenient to
solve these equations in a system of coordinates fixed
with respect to the body. However, there are as yet
no known complete results of the solution of such problems even for simple special cases. All the data considered above were obtained from a solution of the kinetic equations in a coordinate system fixed to the moving body, i.e., from the solution of a number of stationary problems that do not take into account the dependence of the distribution functions f_e and f_i on the
time. However, it is obvious that since the body moves
supersonically, low-frequency (VLF) waves, which depend
on the motion of the ions, must play a role in one form
or another and be manifested in the processes of interaction between the body and the plasma even in the stationary case. Second, it is to be expected that the inhomogeneous cloud carried along by the body, interacting
with the oncoming streams of particles or waves, could
facilitate the processes of excitation or radiation of
oscillations and waves of the plasma in the cloud. At

the same time, it must be borne in mind that because of
the influence of the geomagnetic field all these phenomena take on a specific and more complicated nature.
Therefore, under certain conditions, individual cases
may be distinguished, i.e., the excitation of special
spectra of waves may be favored.

As a whole, this entire problem is a very complicated and little investigated branch of plasma theory
and its systematic study is one of the most immediate
and important problems in the theory of plasma flow
around bodies. The role of wave processes in the neighborhood of rapidly moving bodies is illustrated, in
particular, by the brief review given below of the results of investigations of the wake of a rapidly moving
body.

1. Relationship between the Perturbation $\delta N(r)$ of
the Electron Density of the Wake of a Body and Ion-
Acoustic Waves. The "wave" nature of the perturbation
of the electron density in the far wake of a body is
manifested directly when one considers its Fourier trans-
form δNq (see (2.56)). In the foregoing section we
showed that it is precisely for δNq that one can obtain
(Pitaevskii[60]), with allowance for the influence of the
electric field that is formed in the wake of the body,
a selfconsistent solution of the kinetic equation and
the Poisson equation in the form of closed formulas.
This made it possible to analyze the structure of the
wake of a body. The corresponding results were given
above in §9. For this, we used an inverse Fourier transformation of δNq, i.e., numerical integration of the
formula

$$\delta N(r) = \int \delta Nq \exp(-iq \cdot r) \, d^3q.$$

For simplicity, we illustrate here for an isotropic
plasma, i.e., when $H_0 = 0$, the corresponding properties
of δNq and its "wave" features. In this case, for a
nonisothermal plasma

$$\delta Nq = -N_0 \, (\pi\sqrt{\pi}/q) \, (\rho_0^2 V_0 / v_i) W \left(\frac{V_0}{v_i} \cos \chi\right) \tag{2.69}$$
$$\times \left\{1 + (T_e/T_i) \left[1 + i\sqrt{\pi}\frac{V_0}{v_i} \cos \chi \, W\left(\frac{V_0}{v_i} \cos \chi\right)\right]^{-1}\right\},$$

where, as above (see (2.63)) $W( )$ is the Kramp function
and χ is the angle between V_0 and q. We note that the

denominator of (2.69), if it is equated to zero, gives
the dispersion equation for longitudinal ion-acoustic
waves, $\omega = q \cdot V_0$, when $qD = D/\lambda \ll 1$ (D is the Debye radius).
A formula for δNq and also for the perturbations of the
electric and the magnetic field in the wake of the body
can however be obtained by a method that differs from the
one used in ref. 23. The corresponding calculations,
made in ref. 37 by a method proposed in ref. 60, show
directly that the nature of these perturbations is determined by the excitation of ion-acoustic waves by the
body. Thus, the presence of the dispersion denominator
in (2.69) has a perfectly clear physical meaning, and
the theoretically obtained maximal perturbation of the
electron density on a cone with opening angle $2 \sin^{-1}$
(v_i/V_0) (see Figs. 24 and 25) is the result of C h e r-
e n k o v   e x c i t a t i o n   o f   i o n-a c o u s t i c   w a v e s
by the body. In a certain sense, we can here note an
analogy between this cone and the Mach cone of aero-
dynamics. But we must bear in mind the comments made
above in p. 65-66. In an isothermal plasma, because
of the strong damping of ion-acoustic waves, the cone
of the perturbed region is smeared out and does not
have sharp boundaries (Fig. 24). In a nonisothermal
plasma, in which the damping of ion-acoustic waves is
weak, this effect becomes more pronounced. The Cherenkov excitation of ion-acoustic waves means that the
regions depleted of particles acquire sharp boundaries
and become much narrower -- their angular width decreases
appreciably (see Figs. 27 and 28). The appearance of
accumulation regions behind the body, which with increasing T_e/T_i acquire a narrow "petal" nature (Fig. 28), is
explained, as we have already pointed out in §9, by the
focusing of particles, the enhancement of the influence
of the electric field, and also by the weak damping of
the ion-acoustic waves.

In a magnetized plasma, when $H_0 \neq 0$, the formula
for δNq has a structure similar to (2.69) for an isotropic plasma (see, for example, ref 5). Also, as in
the case $H_0 = 0$, δNq has a dispersion denominator, whose
behavior is determined by the spectra of different types
of magnetoacoustic wave. It is of course harder to investigate the structure of the perturbation when $H_0 \neq 0$
than it is in the isotropic case (see ref. 171, Schmitt).

2. Interaction of Incident Electromagnetic Waves
with the Wake of a Body. It is to be expected that in
the inhomogeneous cloud carried along by the body one
could observe, under certain conditions, various kinds
of resonance due to the interaction of currents excited

in the cloud with the field of the incident electromagnetic waves. Evidently, the oscillations of the plasma
that arise in the wake itself may also lead to a modulation of the incident waves. In the wake, one can expect the simultaneous coherent excitation of oscillations
at frequencies that satisfy different resonance conditions
of the plasma, for example, in the neighborhood of the
upper and the lower hybrid resonances, which will lead
to a complicated structure of the total resultant field.
Various experimental facts observed from satellites,
which will be considered , for example, in Chapter III,
have not hitherto received an adequate theoretical explanation, and these are evidently due to wave processes
that arise because of the interaction of the wake of
the body with the natural electromagnetic radiation incident on the wake from the surrounding plasma. Interesting data of this kind, which require a further theoretical evaluation, were recently communicated, for
example, on the basis of results of observations from
the satellite OGO 1 (Helliwell[64]). A distinctive feature of such experimental facts may be, for example, their
relationship to the period of rotation of the body, the
orientation of its wake with respect to the vector H_0 of
the geomagnetic field, the relationship between the velocity of the waves incident on the body, and the velocity
V_0 of the body itself, and so forth. It must be emphasized once more that this entire group of questions has
not yet been sufficiently investigated (see ref. 55 and
ref. 72 of Sayasov and Zhizhimov). As in the foregoing
subsection, let us, as an illustration, describe briefly
one theoretically considered case of "resonant" interaction of a wave packet incident on the wake of a body;
this treatment evidently explains some experimental data.

It was shown (Bud'ko[65,37]) that when the "resonance"condition

$$d\omega/dk = V_0 \qquad\qquad\qquad (2.70)$$

is satisfied, i.e., when the group velocity of the incident wave packet is equal to the velocity of the body
and also $V_0 \perp H_0$, then in the inhomogeneous wake oscillations are excited corresponding to the resonant branch
of oscillations in the neighborhood of the lower hybrid
frequence ω_L (see (1.34)). Physically, the condition
(2.70) means that the wave packet incident on the body
moves together with the wake of the body, which facilitates a strong interaction between them. The incident
wave packet polarizes the inhomogeneous cloud, and slow

longitudinal waves are excited in the cloud and the effect has long duration. These theoretical calculations
were stimulated by the results of experiments on the
satellite Alouette (Brice, Smith,[66] McEven, Barrington[67]),
in which similar effects were observed. Namely, whistlers, which were detected on the satellite (see branch
a in Fig. 50), excited oscillations of the plasma (branch
b in Fig. 50), which were cut off at the lower hybrid
frequency. The results of theoretical calculations
of this effect,[65] which are shown schematically in Fig.
50 (lower part of the figure), agree fairly well with
the time dependence of the frequency of the packet of
excited oscillations in the neighborhood of the lower
hybrid frequency obtained experimentally. The theoret-

ALOUETTE, 1 MAY 1964

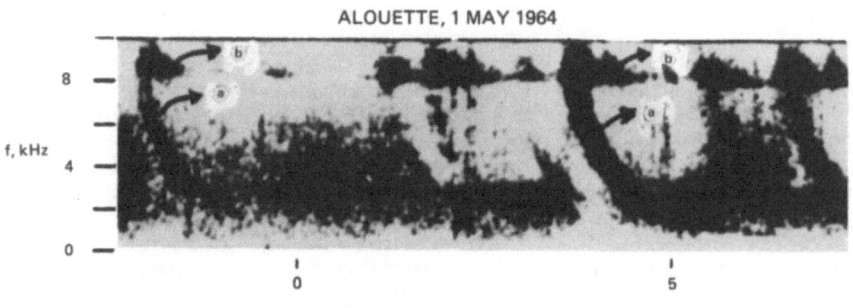

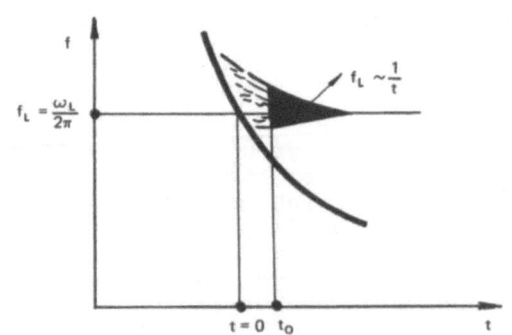

Fig. 50

ical results given in Fig. 50 correspond to times

$$t \gg t_0 \sim \lambda_L / V_0 , \qquad (2.71)$$

for which simple and comprehensible formulas can be obtained. In (2.71), λ_L is the wavelength of the excited oscillations. The width Δf_L of this packet of oscillations and their amplitutde E_L vary as functions of the time as follows:

$$\Delta f_L \sim 1/t ,$$
$$E_L \sim 1/t^{3/2} , \qquad (2.72)$$

For the results of the experiments plotted in Fig. 50 from the data of different measurements on the Alouette satellites (see refs. 60 and 67) $t_0 \sim 0.1$ sec. It should however be noted that a quantitative comparison of the theory with these experimental data could not be made, since the necessary data for this for the amplitude of the oscillations E_L, its dependence on the time, the width of the packet Δf_L of the excited waves, etc, were not given in refs. 66 and 67. At the same time, it is obvious that even this isolated case that has been studied and found to give agreement with the experiment confirms the great interest in the effects that we have discussed in this section.

3. Emission of the Wake of a Body and its Instabilities. We feel it would be desirable to conclude this section with some brief comments on this question. We have seen above that in the neighborhood of the body waves can be excited by the motion of the ions, since the velocity of the body satisfies $V_0 \gg v_i$. If such oscillations are to be weakly damped it is necessary, first, that their wavelength be less than the mean free path of the ions and, second, that the plasma be non-isothermal $(T_e \gg T_i)$. Since the velocity of the electrons satisfies $v_e \gg V_0$, and therefore the phase velocity of the expected electron waves is also greater than V_0, the excitation of electron waves requires more favorable and special conditions than the excitation of ion waves. However, "electron resonances" may be facilitated, for example, by the scattering of electrons on the inhomogeneous (quasiperiodic) structure of the wake of a body, a favorable orientation of the vector H_0 with respect to the vector V_0 and other factors. There are therefore not yet sufficient reasons to exclude the possibility of their excitation. It should be noted that in the experiments noted above on the satellites Gemini-

Agena[10] and Explorer 31 (Samir, Wrenn[166]) it was found
that near the body, behind it on its axis, the effective electron temperature $T_{e,eff}$ was greater than T_e
in the surrounding unperturbed plasma by a factor 1.5-2.
This was evidently due to the fact that the Maxwellian
distribution of the electrons was here destroyed. However, whatever may be the cause of this effect, it could
sustain instabilities of the plasma and the excitation
of waves around the body.

Conditions corresponding to instability of the
plasma and the growth of perturbations in the neighborhood of different types of "resonances" also occur in
the wake of a body. First, the instability associated
with an inhomogeneous distribution of density and temperature of the plasma in the neighborhood of the body.
In the presence of an external field, the perturbation
of the plasma in this case may lead to the growth of
certain kinds of oscillations. Second, in the immediate
proximity of the body the effective ion temperature satisfies $T_i \ll T_e$ (see refs. 5 and 18), which leads to the
formation of two streams of particles, i.e., to a two-
stream instability of the plasma, which facilitates the
growth of small deviations from the stationary state of
the plasma and the excitation of wave processes in the
plasma. It should however be pointed out that under
the real conditions the growth rates in this case are
small.

In problems of this type, it is natural that an
important, and in a number of cases possibly decisive
role, will be played by the interaction between the wake
of the body and the streams of particles -- the corpuscular radiation of the Sun -- incident on the wake. In
this case, as in the study of other problems, the properties of the surface of the body (the nature of its
interaction with the particle streams, see above in §
6) and its potential ϕ_0 (see, for example, ref. 68) are
important. Some questions of the instability of a plasma in the neighborhood of moving bodies, in particular,
those due to the influence of photoemission of its surface, were considered in refs. 173 and 174 (Smirnova).
An interesting form of instability, which can also play
a role in the neighborhood of moving bodies, was investigated in ref. 175 (Gurevich, Pariiskaya, Pitaevskii). This instability mechanism is due to the presence
in the ionosphere of ions of different species, which
leads in the perturbed zone behind the body to the appearance of an electric field that accelerates the motion

of light ions and to the possibility of excitation of ion-acoustic waves.

CHAPTER III

WAVES AND OSCILLATIONS IN THE IONOSPHERE AND NEAR-EARTH PLASMA

§13. Brief Characterization of Results of Different Experiments

The wave processes that take place in the ionosphere, plasmapause, magnetosphere, and also in the transition region (magnetosheath), interplanetary medium, and in the solar wind right out to distances of a million or more kilometers from the Earth have been studied very intensively and successfully in the last few years. Almost every new journal issue (for example, of the Journal of Geophysical Research) carries new experimental data or gives the results of different attempts to explain them theoretically. The state of this branch of plasma physics can be characterized by saying that the experimental results here lead the way, and great progress is being made in different methods of measurements. Although a number of theoretical calculations or solved problems have made it possible to extend our ideas both as regards the wave excitation mechanisms in these regions of the plasma and as regards the propagation properties of different types of wave in the ionosphere, most of the experimental facts remain unexplained and frequently even appear incomprehensible. This is partly due to the fact that the experimental data are frequently inadequate for the theoretical analysis of the results of different experiments, since the quantities characterizing the state of the plasma are not measured simultaneously with the spectra of wave processes. However, the main reason why such a situation has arisen is evidently because certain phenomena cannot be explained in the framework of the linear theory and also because of the great complexity of the kinetic treatment of a number of actual problems; frequently it is also because it is quite unclear how the corresponding theoretical problem should be posed at all.

The treatment of this extensive and many-sided branch of investigations and the generalization of the very large amount of information obtained in many experiments and published in the literature present a difficult task at the present time and, of course, go far beyond the present scope of this book. A separate monograph could and should be devoted to this question. However, in this chapter we should like to give the reader a clear, if brief, picture of the situation with regard to investigations of wave processes in the neighborhood of the Earth. Naturally, this would be impossible unless we restrict ourselves to definite narrow limits. With this in mind, the following plan would appear to be the most sensible and relevant.

First, to present only the basic experimental results and to point out the types of waves that have been observed in a wide range of frequencies in the near-Earth plasma. In doing so, it will of course be necessary for us to make some brief comments of a theoretical nature.

Second, to consider the experimental data obtained directly in the plasma, i.e., basically the results of measurements on artificial satellites and space probes. However, a complete picture of the types of waves observed in the near-Earth plasma cannot be obtained without using some of the results of terrestrial observations. Therefore, some data of this kind will also be presented.

Third, to restrict oneself basically to considering the results of investigations in the topside ionosphere (i.e., out to distances from the center of the Earth of $(4-5)R_0$), in the plasmapause, and in the transition region of the plasma nearer to the Earth in the magnetosphere, i.e., out to distances of $(6-10)R_0$. However, we need not stick rigorously to this restriction. To obtain a complete picture of the observed processes it is worthwhile including various facts obtained at much greater distances, namely, in the solar wind at distances of approximately a million kilometers from the Earth.

In the framework of such a restricted study of the wave processes in the near-Earth plasma, the results of the different experiments considered below can be characterized by the following features.

1. Observations made on artificial satellites and

space probes have revealed plasma waves and oscillations in all the resonance regions predicted by the linear theory (see Fig. 3), namely: the resonance branch adjoining the ion gyroresonance, the branch of oscillations between the lower hybrid frequency and the electron gyroresonance, Langmuir waves, and waves at the upper hybrid frequency. In addition, oscillations peculiar to a nonisothermal plasma ($T_e \gg T_i$) have been observed: longitudinal (electrostatic, $k_0 \| E$) ion-acoustic waves (magnetoacoustic waves) (see Figs. 5 and 6), and multiple ion and electron gyroresonances (see formulas (1.73) and (1.74)).

2. An important and fundamental feature of the results of many experiments is the fact that the observed waves have frequently been detected as transverse electromagnetic waves, i.e., $k_0 \perp E$, H and the data obtained satisfy well the relation n = cE/H, which relates the electric and the magnetic components of transverse waves (n is the refractive index of the wave and c is the velocity of light). At the same time, by their nature these waves must frequently be excited solely as longitudinal waves -- their electric components E must be essentially predominant. This indicates that in the near-Earth plasma longitudinal waves are transformed into transverse waves; hitherto, the actual mechanism has not been elucidated theoretically for any single series of experiments, and this is a very important task of future investigations. This fact was also noted long ago on the basis of numerous results of terrestrial observations, in which one observes only transverse electromagnetic waves from the ionosphere. At the same time, the corresponding electromagnetic oscillations are detected in the frequency ranges in which in the near-Earth plasma resonant branches of waves are excited for which the longitudinal component of the electric field is appreciably greater than the transverse components.

3. Resonances at the lower hybrid frequency are observed particularly often. Quite generally, the lower hybrid frequency plays an important role in the various effects that occur in the near-Earth plasma. At this frequency one observes a cutoff of the oscillations excited in the plasma, and waves propagating in the plasma from sources very far from the point of observation are reflected. In the range of frequencies between the lower hybrid frequency and the ion cyclotron frequency waves are captured in the near-

Earth plasma, since they can propagate at any angle to
the magnetic field. This leads to complicated paths
of very low frequency (VLF) waves. The occurrence of
effects of this type is also facilitated by the multicomponent ion composition of the plasma up to altitudes
$z \approx 1000$ km. Finally, with satellites one can observe
waves that cannot be detected on the surface of the
Earth, namely: subprotonospheric whistlers, magnetospheric-reflected whistlers and their modifications (MR
ν-whistlers), and other effects.

   4. In many experiments one has observed both broad-
band and narrow-band oscillations of the plasma w h o s e
e x c i t a t i o n   m e c h a n i s m   i s   u n k n o w n. Some
results of such experiments are described below. Among
these data, particular interest attaches to the stable
existence for a period of many minutes of narrow-band
longitudinal waves at the frequencies $(3/2)\omega_H, \ldots, \frac{1}{2}(s+$
$1)\omega_H$ (s is a positive integer) predominantly in the neighborhood of the magnetic equator. Observations have
been made of the excitation of oscillations of the second
and third harmonics at $2\omega_U$ and $3\omega_U$ (ω_U is the upper hybrid frequency) and the second harmonic at $2\omega_0$ (ω_0 is
the electron plasma frequency) under the shock effect
of radio wave pulses emitted from satellites. Oscillations of the plasma have been observed at the combi-
nation frequencies $\omega_U - \omega_H$, $\omega_0 - \omega_H$, and also at different frequencies that cannot be identified in a simple
manner with characteristic resonance frequencies known
from linear plasma theory.

   5. Inspection of the substantial literature (which
includes hundreds of papers) describing wave processes
in the near-Earth and interplanetary plasma shows that
there is no established terminology or classification
of different waves according to frequency ranges; this
is partly due to the fact that one does not know at
what altitudes or how these waves are excited. Such
waves, whose frequencies are greater than the gyrofrequency ω_H or the electron plasma frequency ω_0 and are
observed at large distances from the Earth, have in
some papers been called very low frequency or low frequency (VLF or LF) waves. However, by their nature,
these are purely high-frequency (HF) waves, since they
are due solely to oscillations of electrons. In other
papers VLF or LF waves are defined as waves of frequency
$\omega < \omega_L$ (ω_L is the lower hybrid frequency) or frequency
between ω_L and ω_H, i.e., waves whose behavior is essentially determined by the oscillations of ions; it is

when $\omega \gg \omega_L$ and $\omega \to \omega_H$ that the role of ions gradually disappears. In this case, the wave designations are more justified. Similarly, there is no clear definition of the term extra low frequency (ELF) waves. These frequently include waves of frequency $\omega > \Omega_H$ (Ω_H is the ion gyrofrequency) and even of frequency $\omega \gtrsim \omega_L$. In other papers, the adjective ELF is applied to waves with $\omega < \Omega_H$. In Russian literature the terminology especially low frequency waves has been introduced; it is used in a quite obscure manner.

On the basis of a more detailed acquaintance with the multitude of experimental data, I feel it is necessary to introduce a more precise classification of the waves and oscillations observed in the near-Earth plasma according to frequency ranges. The absence of such a classification leads to confusion. It appears to me physically justified to use the classification mentioned above to describe the branches of resonance oscillations (1.31)-(1.35). Accordingly, we introduce the following system:

ELF: waves in the frequency range $0 < \omega \lesssim \Omega_H$
VLF: waves in the frequency range $\Omega_H \lesssim \omega \lesssim \omega_L$
 LF: waves in the frequency range $\omega_L < \omega \lesssim \omega_H$
 HF: waves in the frequency range $\omega > \omega_H$.

Of course, any terminology and classification contains arbitrary elements. In the given case it is also difficult and infrequently even fundamentally impossible to draw a boundary between the different classes of waves. It is particularly difficult when one is considering waves whose mechanisms are due to the nonisothermal state of the plasma. In this case, the chosen boundary ω_L between the VLF and LF waves loses its meaning, since the characteristic frequency in this range of frequencies is not the lower hybrid frequency ω_L but the ion plasma frequency Ω_0.

It is self evident that in each of the following subsections we shall have to include data that overlap with the following subsection. However, the classification we have adopted enables us to make a more definite physical approach to the various experimental data and to set forth the corresponding results in a more elegant manner.

§14. Results of Investigations of ELF Waves $(0 < \omega \leqslant \Omega_H(H^+))$

Here we include in the ELF range only those wave processes whose frequency is below or of the order of the proton gyrofrequency $\Omega_H(H^+)$. In the region of the ionosphere in which different ion species play a role, we also consider certain phenomena due to the behavior of the waves in the frequency ranges between the gyroresonances of the individual ions (see Fig. 4 and its description). The literature data show that the smallest number of experimental data corresponds to precisely this frequency range. As far as I have been able to make out from the results of different experiments in a number of papers in which the observed oscillations of the plasma are called ELF noise or waves, the corresponding processes were excited in regions of the plasma where their frequency appreciably exceeded $\Omega_H(H^+)$. For example, in ref. 69 (Russell, Holzer) there is a report of observation of what the authors call ELF noise in the frequency range f = 100-800 c/s, predominantly at distances from the Earth of $R \sim 3.5R_0$ (L $\approx$ 6) and magnetic latitude ~45°, where $\Omega_H(H^+) \leqslant 100$ c/s. Thus, the data obtained in these experiments may refer to VLF waves, since the frequency range corresponding to them lies in the region $\Omega_H < \omega < \omega_L$. Therefore, if we use such experimental results below, they will be described basically in the corresponding section. In the given case, some of the results of ref. 69 are considered in §15.

1. Hydromagnetic Whistlers. Beginning at the end of the sixties (Benioff,[70] Troitskaya,[71] Saito[72]), observations were made on the Earth's surface of trains of discrete wave packets of magnetospheric origin with frequencies of a fraction of a hertz or a few hertz. They were originally called pearl-type micropulsations Subsequently, it became clear that these wave packets propagate in the electron whistler mode, being guided along the magnetic lines of force of the Earth that intersect the position of the source (Tepley[73]). These signals were therefore called hydromagnetic whistlers. They correspond to the ion wave branch that goes over in the limit $\omega \to 0$ into a slow magnetoacoustic Alfvén wave, and in the limit $\omega \to \Omega_H$ into an ion cyclotron wave (see Fig. 1 and formulas (1.36), (1.37), (1.43) and (1.106)). Having been reflected at magnetically conjugate points, these waves produce a train of discrete signals at the point of observation. The time dependence $d\omega/dt$ of the frequency of these wave packets is

determined by the dispersion law dn/dω. Originally,
wave packets were observed whose frequency increased
with the time: dω/dt > 0 (Tepley, Wentworth;[74] Gendrin,
Stefant;[75] Mainstone, McNicol;[76] Jacobs, Watanabe;[77]
Cambell, Stilner[78]). Examples of sonograms (dependence
of the frequency ω on the time of propagation t) of
these wave packets are given in Figs. 51 and 52 (Kenney,
Knafich;[79] Liemohn[80]). In the upper part of Fig. 51,
we show sonograms of rarely observed wave packets with
the opposite law of time dependence: dω/dt < 0. They
correspond to the electron wave branch that goes over
in the limit ω → 0 into a fast modified Alfvén wave and
for ω > Ω_H and ω_L into an electron whistler (see Fig. 1
and formulas (1.38)-(1.42) and (1.106)).

Figures 51 and 52 differ in a number of respects.
We should first of all point out that it was established
in subsequent experiments that these wave packets are
excited predominantly at distances of (5-9)R_0 from the
center of the Earth (R_0 is the radius of the Earth),
where the electron and ion densities vary approximately
in the ranges 5-20 and 0.1-1 cm^{-3} (Kenney, Knafich, Liemohn;[81] Higuchi, Jacobs[82]). The results of a number of

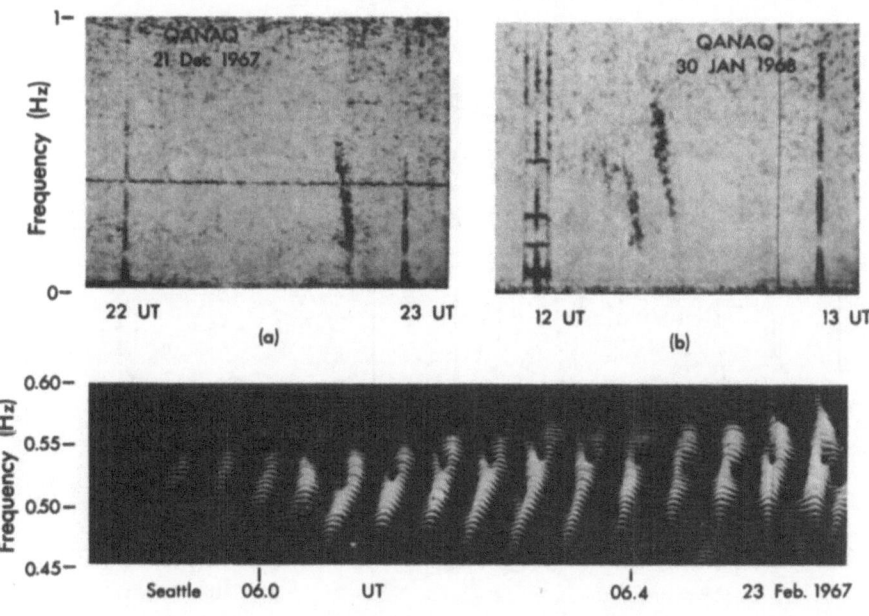

Fig. 51

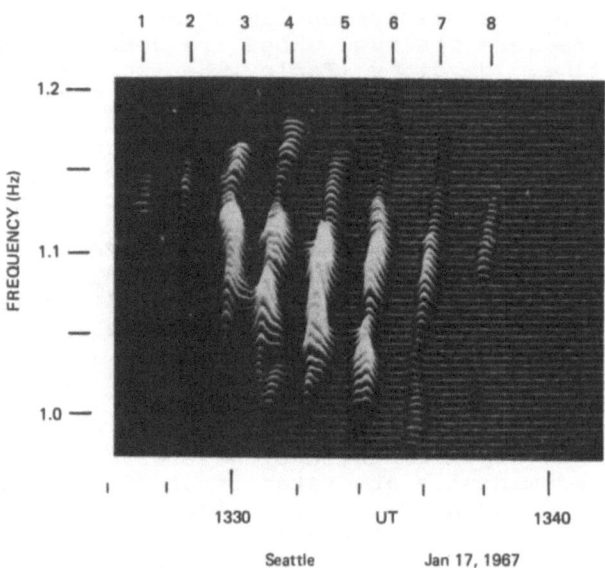

Seattle   Jan 17, 1967

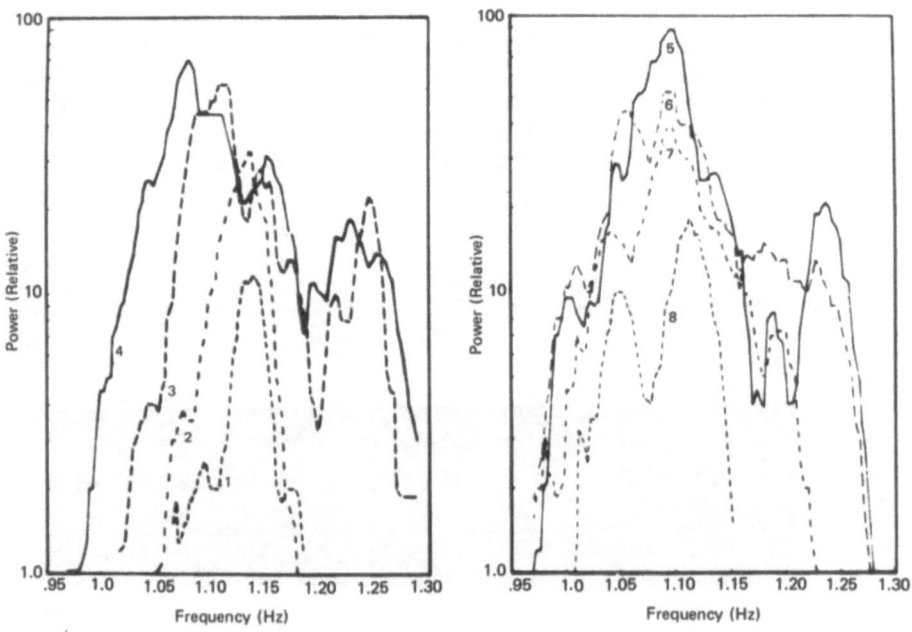

Fig. 52

experiments led to the conclusion that these waves are
excited as a result of ion-cyclotron resonance due to
the interaction between the plasma and streams of particles that encounter the Earth (see formula (1.21)).
Such an excitation mechanism gives good agreement with
the energies of the corpuscular fluxes of $\epsilon \gtrsim 10$ keV.
With regard to the "electron" whistlers shown in the
upper part of Fig. 51, their excitation mechanism re-
mains obscure, since in the corresponding frequency range
$(\omega \sim \Omega_U)$ the electron wave does not have singularities.
If this were an electron cyclotron resonance, the generation of these waves would require quasirelativistic
electrons.

These ideas concerning the ion hydromagnetic whistlers make the sonograms shown in Figs. 51 and 52 more
comprehensible. The figures reveal the enhancement and
not attenuation of the intensity of the signals with increasing signal number. Figure 52 shows the energy frequency spectrum of eight successive hydromagnetic whistlers. Their intensity increases from the first to the
fifth signal and then decreases. The eighth (i.e., last)
signal was, with the first signal, the weakest. It is
assumed that, in addition to the ion cyclotron excitation mechanism, these signals are subject to the ion
cyclotron amplification mechanism when their paths pass
through apogee.[80]

Theoretical calculations of the group delay time
$t(\omega) = \int (1/u_1) ds$ (ds is the element of path length, and
u_1 is the group velocity (see (1.106)) and the determination on the basis of such calculations of the fre-
quency-time characteristics $\omega(t)$ of the hydromagnetic
whistlers (Obayashi;[83] Hultquist[84]) in general agrees
well with the results of the experiment. The corresponding theoretical analysis of the results of the
measurements can be used to diagnose the regions of
plasma in which the signals are generated (see, for example, refs. 81 and 82). At the present time, observations of hydromagnetic whistlers and pearl-type micropulsations generally are carried on by a large network
of stations (Liemohn[85]). Because of the lack of space,
we cannot here dwell on a number of properties of hydromagnetic whistlers, so we refer the reader to the literature data (see also, for example, Pope[86] and Tartaglia[87]).

Hydromagnetic whistlers are excited in regions of the
plasma where one can ignore collisions between the par-

ticles.   Both the mechanism of excitation and amplification and a number of effects that explain the behavior
are of a purely kinetic nature.   The kinetic correction
δ to the refractive index n here plays a small role.
With allowance for thermal motion of the ions,

$$n_{12}^2 = n_0^2 (1 + \delta), \tag{3.1}$$

where n_0^2 is given by formulas (1.36) and (1.38), respectively, for the two branches of waves (1) and (2),
and

$$\delta = \tfrac{1}{2} v_i^2 w \Omega_0^2 / c^2 (\Omega_H - w)^3, \tag{3.2}$$

if

$$z_{Hi} = [(\Omega_H - \omega)/\omega] \cdot (c/n v_i) \gg 1. \tag{3.3}$$

(For more general formulas see refs. 3 and 12).   In ref.
82, the ion density N was determined on the basis of a
theoretical analysis of hydromagnetic whistlers by means
of a formula with a kinetic correction that also takes into account the influence of the loss cone of the charged
particles in the region in which they are reflected.
The results that were obtained show that allowance for
the kinetic correction has little influence on the values of N.   The spatial coefficient of cyclotron damping
of these waves when the condition (3.3) is satisfied and
θ, the angle between the vector k_0 and the vector H_0
of the magnetic field, is equal to zero is

$$\kappa_{Hi} = \tfrac{1}{2} [\sqrt{\pi} c \Omega_H (\Omega_U - \omega)/v_i w^2] \exp(-z_{Hi}^2). \tag{3.4}$$

As $\omega \to \Omega_H$, the value of κ_{Hi} rapidly increases and as long
as

$$[(\Omega_H - \omega)/\omega]^{3/2} \lesssim v_i/V_A \tag{3.5}$$

(V_A is the Alfvén velocity)

$$\kappa_{Hi} \sim c/(V_A^2 v_i)^{1/3}. \tag{3.6}$$

(See refs. 3 and 12 for more detail).

   2. Ion Cyclotron Whistlers.   In the ionosphere the
ion branch of ELF waves was first observed directly from
a satellite in the neighborhood of the proton gyroresonance (Smith et al[88]).   The source of such "proton whistlers" is ELF waves emitted by lightning discharge.

When they enter the ionosphere, these waves are split
into ordinary and extraordinary components. One of them
-- the extraordinary wave -- forms an electron whistler
(see Fig. 1). It is well known that this wave is guided
along the magnetic field of the Earth and is observed
at the magnetically conjugate point or, having been reflected in the neighborhood of this point, at the point
of observation as a whistler with falling dependence
$\omega(t)$: $d\omega/dt < 0$. Many studies of whistlers have been
made (see, for example, Helliwell,[89] Gendrin[90]). We
shall touch on them briefly below in the description
of VLF and LF waves only in connection with the study
of certain phenomena in the ionosphere. The second
component -- the ion wave (ordinary) -- can be observed
only directly in the ionosphere above the source of the
emission, since it is cut off at the gyrofrequencies of
the various ions, whose values decrease with increasing
altitude. It does not therefore reach the magnetically
conjugate point and is not observed on the Earth's surface. One of the sonograms of ion-proton whistlers,
which were first detected on the satellite Injun 3, is
shown in Fig. 53 (Gurnett, Brice[91]). In the same figure one can see the tail of the electron whistler (short
fractional-hop whistler), which because of the short
path of propagation from the Earth's surface to the
satellite suffers little dispersion. In contrast, a
proton whistler undergoes strong dispersion as ω approaches $\Omega_H(H^+)$ and is usually spread out over a few seconds. Subsequently, helium whistlers were also discovered (Fig. 54, Barrington, Belrose[92]). As yet, whistlers of heavier ions (O_2^+, N_2^+, etc) have not yet been
recorded. They are cut off in the region of very low
frequencies (tens or a few hertz) and cannot therefore
be readily resolved by the present day instruments. It
is interesting that three ion whistlers were observed
simultaneously on the satellite Injun 5: proton, helium,
and a third, whose frequency corresponds to a mass equal
to eight units of the proton mass (Gurnett, Rodriguez[93]).
It is assumed that this could be an ion whistler formed
by the doubly ionized oxygen atom O^{++} or a singly ionized helium molecule He_2^+.

Ion cyclotron whistlers (mostly proton whistlers)
were studied in detail in refs. 94 and 95 (Gurnett,
Shawhan, Brice, Smith;[94] Shawhan[95]). The theoretical
analysis of the available experimental data enabled
these authors to develop a method of diagnostics of the
topside ionosphere. This employs, first, the crossover
frequencies of the tail of the electron and the ion

INJUN III. March 5, 1963, $z_s \approx 900$ km

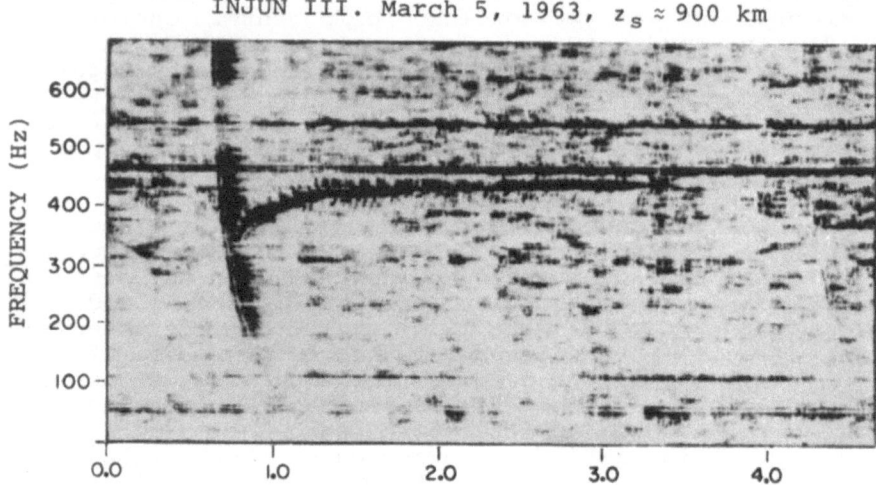

Fig. 53

ALOUETTE II, OCTOBER 4 1966, $z_s = 1190$ km

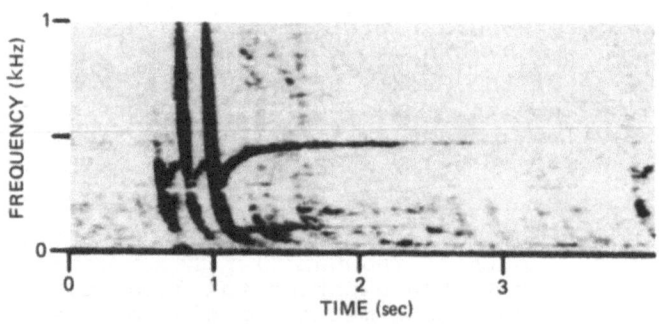

Fig. 54

whistlers (see Figs. 53 and 54 and the frequencies ω_{12} and ω_{23} in Fig. 4), which are determined by the equation

$$n_1^1 = n_2^2 \tag{3.7}$$

by means of formulas (1.43). Second, one determines the dependence of the signal amplitude on the frequency ω up to $\omega \sim \Omega_H$, where the signal is cut off. This enables one to determine the densities of ions and electrons, the proton gyrofrequency, and also the temperature of the plasma in the neighborhood of the satellite (Shawhan, Gurnett;[96] Gurnett, Shawhan;[97] Gurnett, Brice[98]). In addition to the dependence of the relative density of the ions and electrons on the crossover frequencies, the formulas that determine these plasma parameters are obtained from a calculation of the group delay time $t(\omega)$ and the spatial damping β of the signal amplitude, namely, from the integrals

$$t(\omega) = \int (1/u_1)\,ds$$
$$\beta \approx \int \exp\,(-\omega\kappa/c)\,ds. \tag{3.8}$$

In (3.8), u_1 is the group velocity (see (1.104) and (1.106)). As Gurnett and Brice show,[98] the damping β, as a function of the frequency difference $\Delta\omega = \Omega_H(H^+) - \omega$ or the time t, nearly satisfies the experimental results if $\kappa = \kappa_{Hi}$, i.e., if one uses the formulas of ion cyclotron damping (see (3.4)). This proves the kinetic nature of the damping of the proton whistlers. In particular, the corresponding results are illustrated in Fig. 55, which is constructed on the basis of data taken from ref. 98. In the upper part of Fig. 55 we have plotted theoretical curves for the relative amplitude as a function of time, calculated without allowance for damping with allowance for collisions, and using the coefficient of cyclotron damping κ_{Hi}. In the lower part of Fig. 55, we have plotted the experimental dependence of the amplitude of the proton whistler on $\Delta\omega$ (points) and the expected relative value of the amplitude without allowance for damping. In one of the more recent papers (Lucas, Brice[99]) it is shown that the experimentally observed damping of proton whistlers as a function of the time (or $\Delta\omega$) occurs more slowly than one would expect from the calculation of the theoretical dependence with allowance for only cyclotron damping. Lucas and Brice[99] considered the various factors that could lead to this discrepancy and came to the conclusion that the most probable reason is the influence of irregular (cloud-like) structure of the proton density $N(H^+)$ in

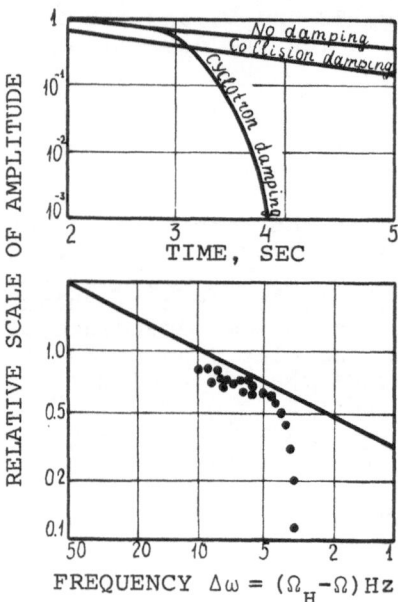

Fig. 55

the neighborhood of the satellite. They showed that with allowance for the fluctuations δN the temperature obtained from the analysis of the proton whistlers may even be approximately twice the temperature determined on the basis of the formulas without allowance for δN (see ref. 98). We also mention here that, as was shown recently in a theoretical paper (Stefant[111]), proton whistlers can have a fine structure due to the influence of gyroresonances of the oxygen ions O_I^+. Stefant[111] points out that in Fig. 9 in ref. 108 a corresponding fine structure -- modulation of proton whistlers -- can be noted.

    3. ELF Hiss and Chorus. Cutoff and Amplification of ELF Waves as $n \to 0$ between the Proton and Helium Gyrofrequencies. In the topside ionosphere, as also on the surface of the Earth, one observes continuous emission in discrete frequency ranges with the nature of white noise: this is known as hiss emission. In addition,

again as on the Earth's surface, one also observes chorus emissions which consist of sequences of randomly
formed discrete signals with a duration of a few tenths
of a second; their frequency usually increases with the
time. An example of a sonogram of this type of emission
adjoining the ELF frequency range, which was obtained
from observations with the satellite Injun 3, is shown
in Fig. 56 (Taylor, Gurnett[100]). The frequency of the
ELF emissions, which Taylor and Gurnett called ELF hiss
and chorus, varied from a few hundred hertz to approximately 2 kHz. From a first glance at this data it would
seem that the lower part of this frequency range corresponds to frequencies $\omega < \Omega_H(H^+)$, i.e., that it lies in
the range of frequencies less than the proton gyrofrequency; in the ionosphere, the maximal value of this
frequency is 600-650 Hz. The upper boundary of this
emission lies at $\omega > \Omega_H(H^+)$ and goes over into the region of VLF waves. However, a number of subsequent investigations have shown that this emission in fact consists of VLF waves and is only cut off in the ELF frequency range. The emission is generated at higher altitudes in the topside ionosphere or in the magnetosphere
at frequencies ω greater than the gyrofrequency $\Omega_H(H^+)$
of the corresponding region of plasma. This follows
from the following data.

It was shown in experiments on Injun 3 that emission of the ELF hiss and chorus type is often sharply

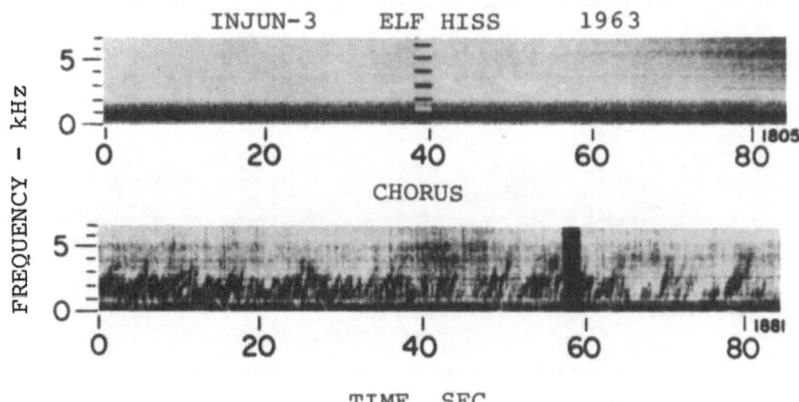

Fig. 56

cut off at a given altitude at a frequency $\omega = \omega_c$ between the proton and helium gyrofrequencies at which $n_1^2 \to 0$ (in Fig. 4 this corresponds to a value $n_1^2 = 0$ between Ω_{H1} and Ω_{H2}). Such results are shown, for example in Fig. 57 for hiss type emission and in Fig. 58 for chorus type emission (Gurnett, Burns[101]). It can be seen from these figures that with decreasing altitude the frequency at which the emission is cut off increases, i.e., the gyrofrequencies $\Omega_H(H^+)$ and $\Omega_H(H\overset{+}{e})$ increase. On the basis of these data, Gurnett and Burns[101] concluded that this emission was excited above the satellite and propagated originally as a packet of extraordinary waves (the branch n_2^2 in Fig. 4). At the altitude at which the frequency ω becomes equal to the crossover frequency ω_{12} the polarization sense of the extraordinary wave changes. Therefore the subsequent propagation of waves of frequency $\omega < \omega_{12}$ is described by the branch n_1^2 and they are cut off at the frequency ω_c, at which $n_1^2 = 0$. Thus, below the altitude at which $n_1^2(\omega_c) = 0$ only part of the wave spectrum arriving from higher can propagate, namely, waves with frequency $\omega < \omega_c$. It follows from Fig. 57 that at $z \sim 1000$ km we have $f_c \sim 400$ Hz, and there are therefore reasons for believing that in the considerd case the lower boundary of the spectrum of the observed hiss was indeed less than the minimal values of ω at which these waves were observed at $z \sim 1000$ km. In subsequent experiments on Injun 5 it was shown by direct measurements of the Poynting vector that this type of signal does indeed propagate predominantly downward (Mosier, Gurnett;[102] Mosier[103]).

INJUN 3. 1963

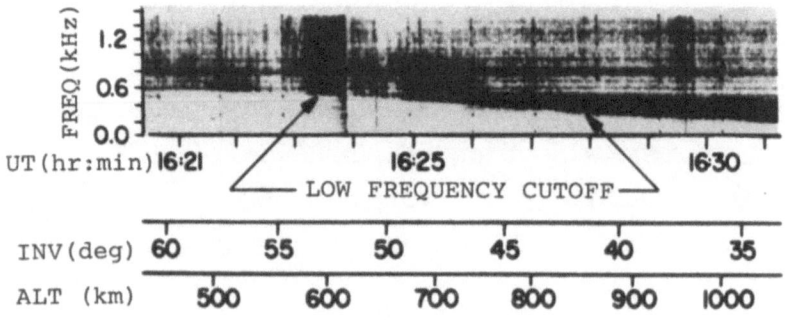

Fig. 57

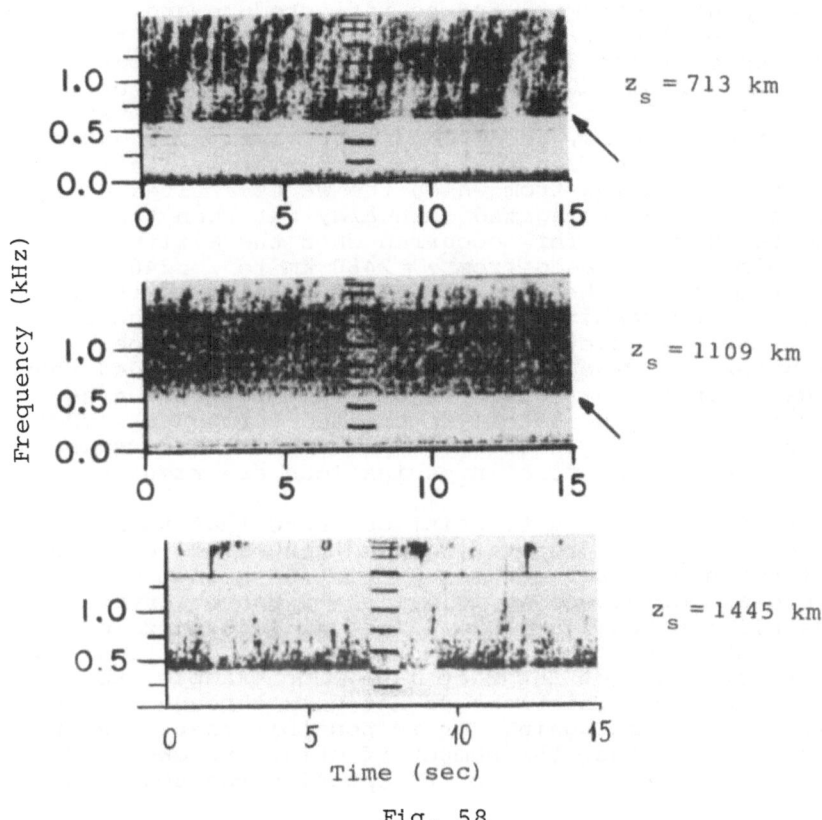

Fig. 58

However, in a narrow range of frequencies at the cutoff
limit of the signals there were also observed less
strong waves that propagated upward, i.e., had been reflected from the lower lying regions of the ionosphere.
This is evidently explained by the fact that in some
parts of the ionosphere below the satellite the cutoff

frequency was lower and not higher than the value of ω_c
near the satellite, so that waves with frequency $\omega < \omega_c$
could propagate downward and be reflected upward. Interesting results of observations of mixing of ELF hiss
and chorus on Injun 5 are shown in the color sonogram
in the range from 250 Hz to 1 kHz (Fig. 59) (Gurnett,
Mosier, Anderson[104]). The red color indicates waves
propagating downward, the green color waves propagating
upward. It can be seen that at approximately 14h 18 min
40 sec the direction from which the waves arrived changed.
Originally, they propagated downward, but then they
propagated upward. This occurred when the altitude of
the satellite decreased from z = 2480 km to z = 2464 km.
At the same time, the upper and lower limits of the emission frequency remained almost unchanged, although a
more strongly discrete structure -- of chorus type --
was manifested. The possibility cannot be excluded that
the change in the direction of arrival of these waves
occurred because the satellite crossed through the region
in which they were generated. For complete understanding
of this phenomenon, more investigations are required.

It is interesting to point out here that the reflection of upward propagating electron waves (whistlers)
at a frequency ω_c between the proton and helium gyrofrequencies corresponding to $n_{\frac{1}{2}}^2(\omega_c) \to 0$ was observed on
the satellites OGO 2 and OGO 4 (see Fig. 60, Muzzio[105]).
Thus, in the cases when the two modes (the electron and
the ion wave) do not interact (see, for example, Rodriguez, Gurnett[106]) whistlers do not pass through the ionosphere and return again. It is possible that this explains the fact that the number of whistlers observed
on satellites in the hemisphere opposite the source ap-

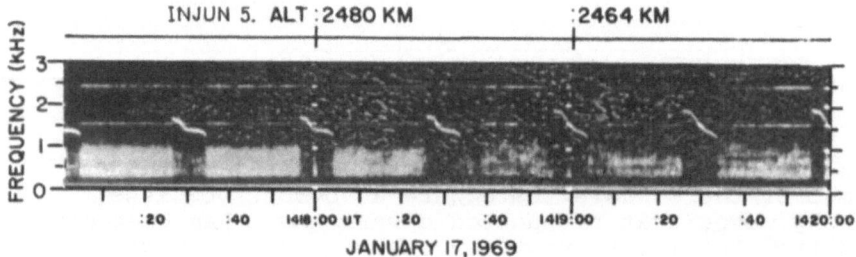

Fig. 59

OGO IV. 3 VIII 1967. $Z_s \approx 900$ km

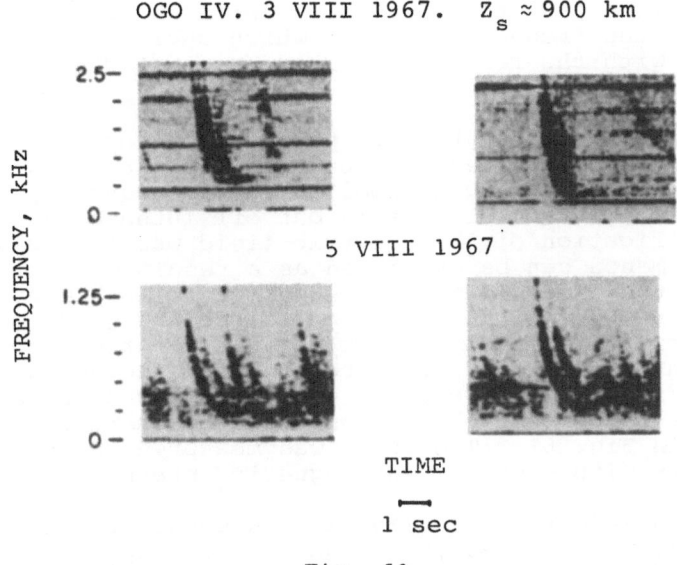

Fig. 60

preciably exceeds the number of whistlers observed simultaneously near the Earth's surface (Gurnett et al[108])

As we have already pointed out above, ELF emission
of hiss type consists of a packet of electron waves
propagating in the electron whistler mode (see Figs. 1
and 4). Simultaneous measurements of their magnetic
and electric components showed that they are indeed transverse waves ($k_0 \perp E, H$), related by the equation that
follows from Maxwell's equation:

$$n = cH/E,$$
$$n = \frac{3 \cdot 10^4 H(Oe)}{E(V/m)} \tag{3.8}$$

(Gurnett, Pfeiffer, Anderson, Mosier, Caufman[108]). In
ref. 108 this was illustrated by an example when for one
of the Injun 5 experiments a value $n \approx 80$ for the refractive index was obtained by means of (3.8) from absolute
measurements of E and H for the frequency ~200-600 Hz.

Independent measurements of the electron density by the
same satellite gave the value $n \approx 95$ with an accuracy
of $\pm 20\%$ at the frequency 400 Hz, which obviously agrees
very well with the results obtained by means of formula
(3.8).

It is of interest here to point out an effect described in ref. 108 that was observed in Injun 5 experiments. In the neighborhood of the frequency $\omega_c <$
$\Omega_H(H^+)$ at which the ELF hiss is cut off intense narrow-
band amplification of the electric field was observed.
This phenomenon can be explained as a result of an increase of the electric field $E \sim 1/n$ (see (3.8)) when
$\omega \to \omega_c$ and $n \to 0$. At the same time, since the group velocity decreases rapidly when $\omega \to \omega_c$, in the region of
reflection of the wave its energy density is increased,
which enhances this effect. A corresponding sonogram
that illustrates this phenomenon (taken from ref. 108)
is shown in Fig. 61. The field was measured in these
experiments with electric and magnetic antennas.

To conclude this section, let us consider briefly
the results of observations of an interesting kind of
narrow band emission that was observed sporadically in
the transition zone of the near-Earth plasma (magnetosheath) at distances from the Earth of $(10-15)R_0$ (Smith,
Holzer, Russell[167]). The same waves have also evidently
been observed at altitudes $z < 3000$ km in the polar ionosphere and have recently been described (Gurnett,
Frank[168]). This emission was called ELF noise, and Smith,
Holzer, and Russell[167] note that, in accordance with
the established tradition, this emission could be called
lion's roar. In accordance with the classificication
adopted in this book, and also in accordance with the physical nature of the propagation and structure of these
waves, as will be seen below, they correspond rather to
the resonance region and properties of VLF or LF waves
(see §3). However, we shall the corresponding results
in this section, adopting the terminology of refs. 167
and 168, one reason for this being that lack of space
prevents our considering the expected excitation mechanisms of these waves.

On the satellite OGO 5, in the transition region
of the interaction of the solar wind with the Earth's
magnetosphere (magnetosheath) above the boundary of the
magnetopause, waves were observed by means of magnetic
antenna in a narrow range of frequencies $f_s \approx 50-200$ Hz
with central frequency ~100 Hz whenever the satellite

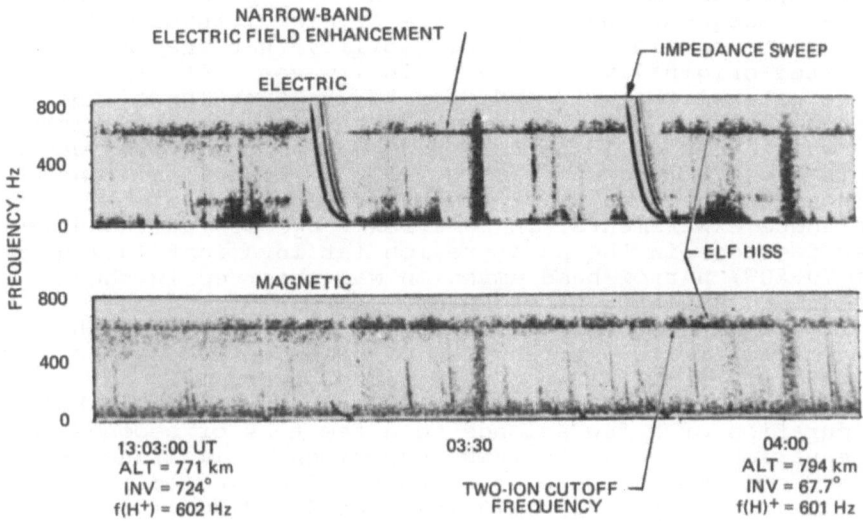

Fig. 61

crossed this region of the plasma. A sample of the corresponding sonogram is shown in Fig. 62. The amplitude
of this emission reached several tenths of a gamma and
the duration varied from fractions of a second to a few
tens of seconds. Smith, Holzer, and Russell[167] assume
that these are transverse electromagnetic whistler-mode
waves propagating in the magnetosheath. Since at these
distances from the Earth the values of $\omega_s = 2\pi f_s$ are
much greater than not only the ion gyrofrequency $\Omega_H(H^+)$
but are also greater than the hybrid frequency $\omega_L \sim \sqrt{\omega_H \Omega_H}$
(see Tables 1 and 2 in §2), these are indeed electron
VLF of even LF waves, and they correspond to the frequency range of the resonance branch $\omega_2(\theta)$ of a cold
plasma (see Figs. 1, 2, and 3) or the branch $\omega_1(\theta)$ of a
nonisothermal plasma (Figs. 5 and 6). However, in these
experiments the electric and magnetic components were
not measured simultaneously and nor was the electron
density. There is therefore no possibility of verifying
by means of (3.8) and formulas (3.21) and (3.22) below
that these are indeed transverse waves. Here we shall
not dwell on the excitation mechanisms of these waves

discussed in ref. 167 or other possible explanations of
this type of emission. We must however point out that
in a subsequent analysis of the origin of these waves
one must bear in mind the possibility that they may be
excited originally as longitudinal waves. It is therefore natural that, in addition to the question of their
generation mechanism, one must also consider mechanisms
by which they could be transformed into transverse waves.
This question becomes particularly important when one
discusses the results of the experiments on Injun 5.
In these experiments, in a narrow range of latitudes (a
few degrees) in the polar region (at invariant latitudes
of 70-80°) narrow-band emission was observed in the
frequency range f_s ~ 100-300 Hz. An example of the
corresponding sonogram is shown in Fig. 62, in which
the magnetic field strength H of these waves is also
given. The band width of the waves observed in ref.
168 generally did not exceed 100 Hz. The emission had
a duration of a few seconds to a few tens of seconds in
the various cases. In these experiments simultaneous
measurements were made of E and H. They varied in the
following limits: E ≈ 3-10 mV/m and H ≈ 10-30 mγ. Gurnett
and Frank,[168] comparing their data with those of ref.
167, concluded that they had observed waves of the lion's
roar type described in ref. 167. These waves are trapped
in the open ends of the lines of the Earth's magnetic
field H_0 that are formed at the upper boundary of the
magnetosphere; the waves propagate along H_0, being
ducted by elongated inhomogeneities, and reach altitudes of z < 3000 km in the polar region. In this case
there is almost purely longitudinal propagation ($\theta \approx 0$),
described by the electron branch n_2 (see Fig. 4).
Therefore, although the frequency f_c of these waves
gradually becomes less than the ion gyrofrequency in the
ionosphere, these waves cannot be cut off when $n_i^2 = 0$ at
frequencies $\omega < \Omega_H$ (Fig. 4). Although in these experiments both field components (E and H) were measured, it
is impossible to prove for certain that they exactly
satisfy the relations (3.8), (3.21) and (3.22) for transverse waves, since there are no data for the electron
density and the magnetic field for the actual cases of
observation. Estimates that can be made on the basis
of published data gives slightly too low values of the
refractive index n.

Undoubtedly, subsequent investigations of this type
of emission in the far regions of the near-Earth plasma
are of great interest, especially if the aim is to
achieve a more certain theoretical explanation of the ef-

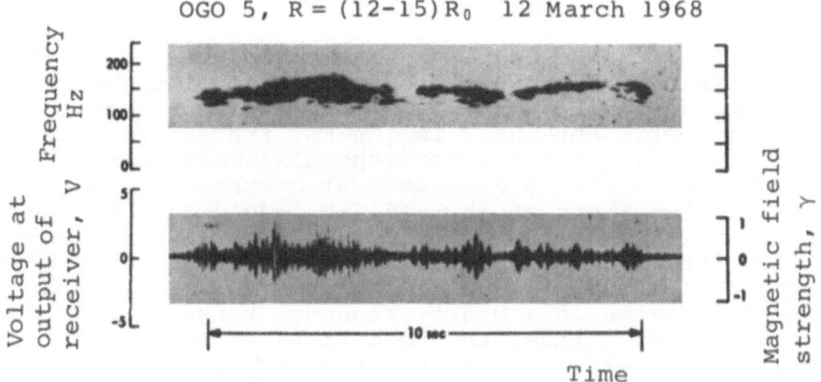

OGO 5, R = (12–15) R₀   12 March 1968

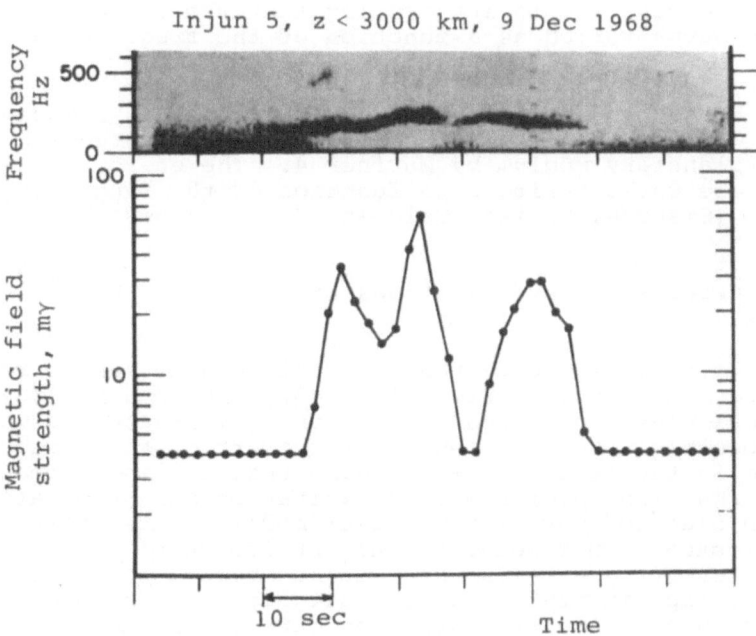

Injun 5, z < 3000 km, 9 Dec 1968

Fig. 62

fect. It would be very interesting to analyze experimentally the fine structure of this emission in the same way as was done in the paper ref. 150, described below in §17, for high frequency waves.

In the magnetosphere and the magnetosheath broadband emission that is assumed to consist of transverse waves propagating in the plasma have also been detected with a magnetic antenna. In the ELF frequency range, namely for $\omega/2\pi \lesssim 0.3$ Hz, corresponding measurements were made on the satellite Pioneer 5 at distances from the center of the Earth of $(5.2-15.4)R_0$ (Coleman[184]). In the experiments of OGO 1 (Smith, Holzer, McLeod, Russell[185]) broad-band emission, which also evidently consisted of transverse waves, was also observed with a magnetic antenna in a wider frequency range, namely for $\omega/2\pi \approx 1-300$ Hz. Under the conditions of these experiments, the lower limit of the frequency range was of order Ω_H, while the upper limit was much higher than the lower hybrid frequency ω_L and even of order of the electron gyrofrequency ω_H. Thus, in these experiments one observes simultaneously a broad band of ELF, VLF, and LF waves. It is interesting that the energy of these waves varied as a function of the frequency as ω^{-3}.

Broad-band emission of only ELF waves at frequencies $\omega/2\pi \approx 3\cdot 10^{-4}-0.5$ Hz $< \Omega_H/2\pi$ was also observed in the interplanetary medium by Mariner 4. The energy density of these waves varied as a function of the frequency as $\omega^{-3/2}$ (Siscose, Davies, Coleman, Smith, Jones[186]).

§15. Results of Investigations of VLF Waves $(\Omega_H(H^+) \leqslant \omega \leqslant \omega_L)$

In this frequency range, a cold plasma has no resonance branches (see Figs. 1 and 3), and in a nonisothermal plasma only the excitation of ion-acoustic waves is possible. Another important feature of this frequency range is the fact that it is here that the influence of different ion species on the excitation and propagation of an electron wave (of whistler mode) is most strongly manifested. This leads to the appearance of a number of interesting effects which it has become possible to detect experimentally and investigate only in recent years in experiments carried on directly in the near-Earth plasma on satellites or rockets. For these reasons, the published experimental data relevant to this section are more varied in nature and physical essence. It is obvious that we shall be able to consider this information only very briefly, almost schematically.

1. Emission at Proton Gyrofrequency Harmonics. Emission Spectra Cut Off at the Proton Gyrofrequency.

In some experiments, wave excitation bands have been observed at the proton gyroresonance harmonics described by equation (1.73) (see §4). These longitudinal waves $(\mathbf{k}_0 \| \mathbf{E})$ can be readily excited, as is well known, when $\mathbf{k}_0 \perp \mathbf{H}_0$, since Landau damping is then weak. Gyroresonance proton waves up to the eighth harmonic were apparently first observed on the rocket Javelin 8.46 (see Fig. 63; Mosier, Gurnett[109] and Gurnett, Mosier[110]). Multiple gyroresonance excitation of protons was also observed on an Injun satellite.[108]

In the experiments on OGO 2 in the range of altitudes 415-1507 km in its orbit, VLF waves were detected that had a maximal amplitude (were excited and cut off) at the proton gyrofrequency $\Omega_H(H^+)$. The frequency range of these waves varied in different observational periods from 300-700 Hz to 18 kHz in approximately the same ranges in which the values of $\Omega_H(H^+)$ and Ω_0, the ion plasma frequency, varied at the same altitude. In these

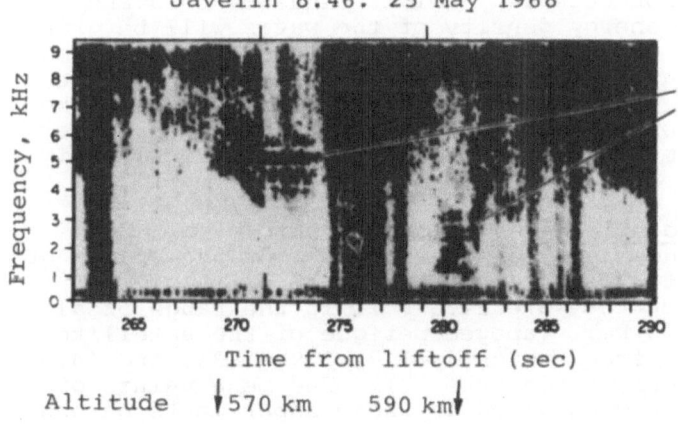

Javelin 8.46. 25 May 1968

Proton gyrofrequency harmonics

Time from liftoff (sec)

Altitude   570 km   590 km

Fig. 63

experiments, only a magnetic antenna was used and the magnetic field strength H was measured (Guthart, Crystal, Ficklin, Blaiz, Yung[112]). A feature of the frequency dependence of the field amplitude in the results that were obtained were its rapid decrease with increasing frequency. The authors of ref. 112 concluded that the emission they had observed consisted of ion-acoustic waves. Evidently, this is the branch of fast ion-acoustic waves excited in a nonisothermal plasma. It was shown in Figs. 5 and 6 (see formulas (1.62) and (1.67)). If this was indeed a longitudinal electrostatic wave, the observed effect was the magnetic field H induced in the antenna by the motion of the antenna with respect to the electric field E. In this case

$$H = \mathbf{V} \cdot \mathbf{E}/c^2,$$

$$H(\text{Oe}) = \frac{VE(\text{cgse})}{c}. \tag{3.9}$$

For a number of measurement runs (at $V_0 \sim 7$ km/sec) the field strength $E \sim 40$ V/m was obtained from (3.9) at the proton gyrofrequency. This gives an energy flux density of the wave of $E^2/8\pi \sim 7 \cdot 10^{-8}$ erg/cm³; this is commensurable with $N\kappa T$ in this region of the ionosphere. Theoretically, the maximally expected intensity of ion-acoustic waves is of order $\frac{1}{2}N\kappa T_e$ (Rostoker;[113] Scarf, Crook, Fredericks[114]). It would therefore seem justified to interpret the results of these experiments as above. It is true that the maximal value obtained for the field strength E still appears too high, since the detected emission occupies a wide frequency band and the integrated energy density of the waves will be greater than $7 \cdot 10^{-8}$ erg/cm³. Of course, since the density N and the temperature T_e of the plasma were not determined simultaneously in these experiments, it is hard to make a final decision as to the extent to which the results of the measurements disagree with the theoretical estimates.

2. Ion-Acoustic Waves. Ion-acoustic waves, described like the waves considered above by the branch of fast ion-acoustic waves (see Fig. 5; $\Omega_H < \omega < \Omega_0$) were also observed on the satellite P 11 in the range of altitudes 268-3720 km; (apogee-perigee of the satellite) at the discrete frequencies f = 1.7, 3.9, 7.35, and 14.5 kHz (Scarf, Crook, Fredericks[114]). The mean values of the electric field obtained in these experiments on one of the orbits is shown in Fig. 64. The results of the experiments are summarized as follows in ref. 114:

Mean level of field .... $E \approx 1\text{-}2$ mV/m,
Occasionally observed .... $E_{min} < 600\text{-}800$ μV/m,
Frequent field bursts on Earth's
nightside with duration 3-10 min .. $E \approx 20\text{-}100$ mV/m
Sometimes observed .... $E \sim 1$ V/m (3.10)

Since the lower limit of the frequency range used
in ref. 114 was $\omega > \Omega_H(H^+)$ or $\omega \gg \Omega_H(H^+)$, it was not possible to observe the cyclotron excitation of the branch of
fast ion-acoustic waves, in contrast to the experiments
described above in ref. 112. Naturally, the field
strength E did not reach very large values, although $E \sim$
1 V/m (see (3.10)) is already a fairly strong field.
The energy density of the waves varied generally in the
range $10^{-16}\text{-}10^{-13}$ erg/cm^3, which is several orders of
magnitude less than $N\kappa T_e$ in this region of the topside

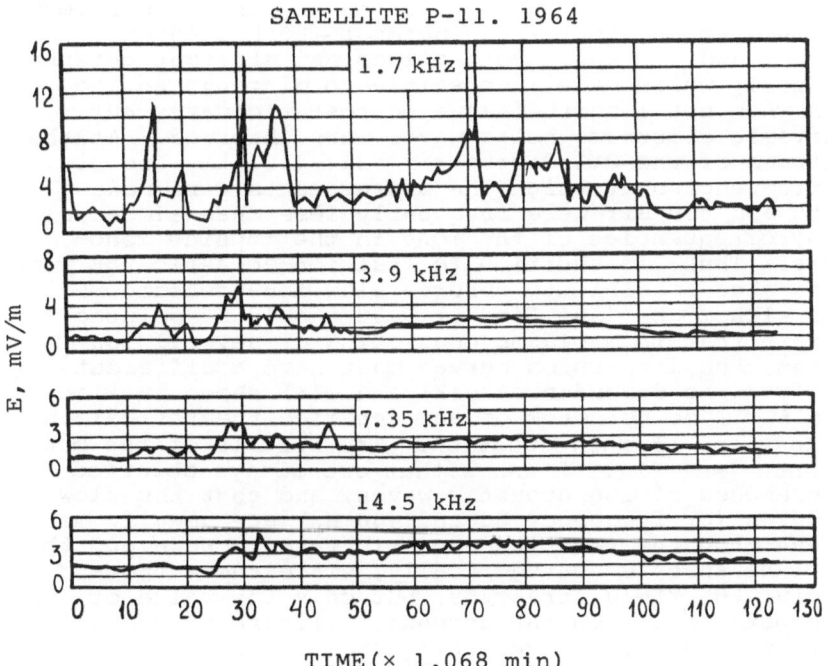

Fig. 64

ionosphere. It should be noted that the minimal field values, E_{min}, observed in these experiments generally agreed well with the theoretical estimates. For example, in accordance with ref. 113 for $\omega \ll \Omega_0$ and receiver band width $\Delta f/f \ll 1$

$$E_{min}^2 = \kappa T_e (2\pi/D_i)(4\pi^2/\Lambda^2)(\Delta f/f) ,$$

where D_i is the Debye radius of the ions and Λ is the wavelength. At the frequency $f = 1.7$ kHz and $\Lambda \approx 90$ cm (such an estimate is given in ref. 114) one obtains $E_{min} \approx 360$ mV/m from (3.10) for the mean values of the parameters of the ionospheric region in which the satellite orbit passed. This value of E_{min} is less than the observed values, which is possibly a further proof of the ion-acoustic nature of these waves. The authors of ref. 114 arrived at the same conclusion in their subsequent experiments with the satellite OV3-3 (Scarf, Crook, and Fredericks[115]), in which technical improvements were made. In these experiments, the measurements were made at the four frequencies $f = 80$ Hz, 400 Hz, 1.65 kHz and 7.3 kHz. Both a loop and straight antenna were used, which made it possible to distinguish the cases when not longitudinal electrostatic waves but transverse electromagnetic waves were observed. At the altitudes of the OV3-3 orbit, $z = 354-4460$ km, two of the frequencies at which the measurements were made ($f = 80$ and 400 Hz) were frequently less than almost all the gyrofrequencies of the ions in the topside ionosphere. Thus, in addition to fast ion-acoustic VLF waves, it was also possible to observe slow ion-acoustic ELF waves with $\omega < \Omega_H$ (see Fig. 6). It should be pointed out here that in a plasma consisting of several ion species, the dispersion curves must have a different form from the dependences $\omega(k)$ and $\omega(\theta)$ shown in Figs. 5 and 6. I do not know of any detailed theoretical investigations of this question. The results of the measurements (see Fig. 65) show that one always observes two branches of ion-acoustic waves, and that the slow ELF waves (of frequency 80 and 400 Hz) are usually more intense than the fast ion-acoustic waves. However, this cannot be asserted categorically, since the authors do not give the field strengths, but only the value of the field potential ε on the antenna. (Figure 65 shows the limits within which ε varies.)

Ion-acoustic waves have also been observed at lower frequencies than in the experiments considered above.

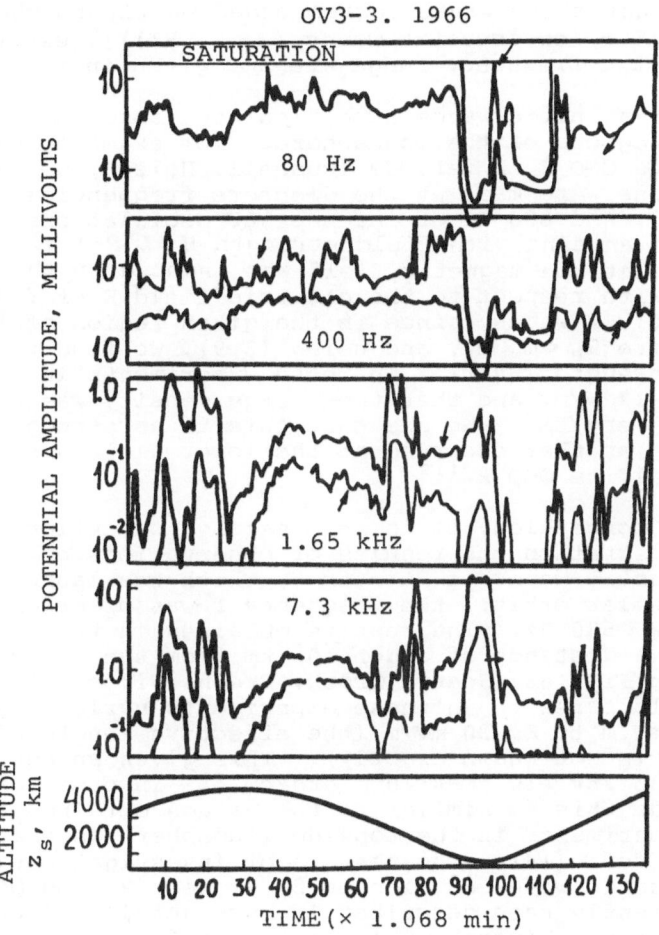

Fig. 65

For example, on the rocket Javelin 8.45 (Shawhan, Gurnett[116]) intense longitudinal waves (large values of E
and absence of an effect on the magnetic antenna) were
observed in the frequency range f < 1 kHz. The amplitude of the electric field E had maximal values at the
minimal altitudes at which the measurements were still
made (z ~ 250-280 km). The possibility cannot be ex-

cluded that these waves corresponded partly to the branch of slow ELF ion-acoustic waves ($\omega < \Omega_H(H^+)$); exact data on the frequency range are not given in ref. 116.

Ion-acoustic waves have also been observed in the higher regions of the ionosphere. For example, on the satellite OGO 3 in ref. 69 (Russell, Holzer, Smith) observations were made at the discrete frequencies f = 100, 300, and 800 Hz of intense VLF oscillations on the magnetic antenna with field strength H $\approx 0.2-1$ γ. Assuming that the magnetic field was induced in the antenna moving with respect to the electric field E of fast ion-acoustic VLF waves, since in the given region of the ionosphere $\Omega_H < \omega < \omega_L$, and using (3.9), we find that the field strength E varied in these experiments in the range 3-12 V/m, and that the energy density was $E^2/8\pi$ ~ $2 \cdot 10^{-9}$ erg/cm^3. We see that this is as strong an emission as that observed at the lower altitudes on the satellite OGO 2.[112]

In conclusion, it is interesting to mention here for illustration the results of observations of fast ion-acoustic waves on Pioneer 8, which was launched into a solar orbit; the frequency f was of order 400 Hz, below $\Omega_0 \approx 530$ Hz. The results obtained in the solar wind at a distance of order 10^6 km from the Earth are shown in Fig. 66 (Scarf, Crook, Green, Virobik[117]). The field strength in these experiments varied from E ≈ 0.2 mV/m to E ≈ 30 mV/m (the effective length of the antenna is not known exactly). This gives an energy density $E^2/8\pi$ ~ $10^{-16}-4 \cdot 10^{-14}$ erg/cm^{-3}; in order of magnitude this is similar to the values obtained in some experiments in the topside ionosphere (see (3.10) above; ref. 114). The results of investigations of ion-acoustic waves on Pioneer 8, Pioneer 9, and OGO 5 have recently been described in more detail (Siscose, Scarf, Green, Binsack, Bridge;[163] Scarf, Fredericks, Green[164]).

3. Excitation of Waves at the Lower Hybrid Frequency ω_L.
At the end of the foregoing section we already gave the results of observations of resonance excitation of oscillations in the ionosphere at the lower hybrid frequency in the neighborhood of satellites. These oscillations or waves were triggered by whistlers (see Fig. 50), and were first observed on the satellite Alouette 1 (Barrington, Belrose;[118] Barrington, Belrose, Keeley[119]). Subsequently, the results of these measurements were used to determine the effective mass M_{eff}

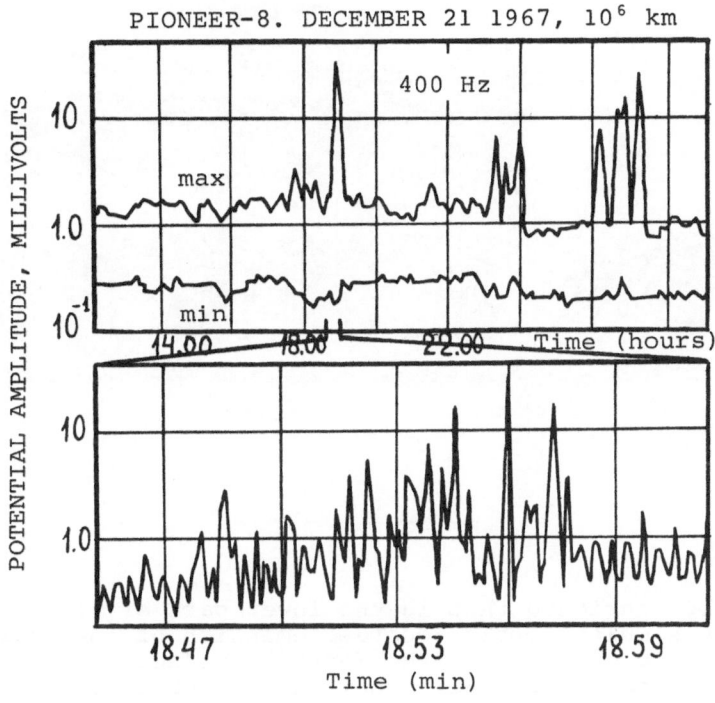

Fig. 66

of the ions (see equation (1.42)) and other parameters
of the ionosphere (Barrington, Belrose, Nelms[120]). In
addition, broad-band emission of the hiss type cut off
at the lower hybrid frequency was also observed on Alouette 1 (Brice, Smith[121]), this being observed predominantly on the electric antenna, i.e., it was a lon-
gitudinal wave $(\mathbf{k}_0 \| \mathbf{E})$. The spectrum of such waves is
shown in Fig. 67 (see ref. 121). It can be seen that
the emission is cut off at the frequency ω_L, whose value
decreased with the time from $f \approx 10$ kHz to $f \approx 5$ kHz, since
the satellite moved to higher latitudes, its altitude
remaining almost unchanged ($z \approx 1000$ km), which brought
it into an area of weaker magnetic field H_0, which therefore reduced ω_L. The possibility cannot be excluded
that these waves correspond to the resonance branch of
LF waves excited in a cold plasma (see Fig. 3). It

ALOUETTE I, 23 OCT 1963

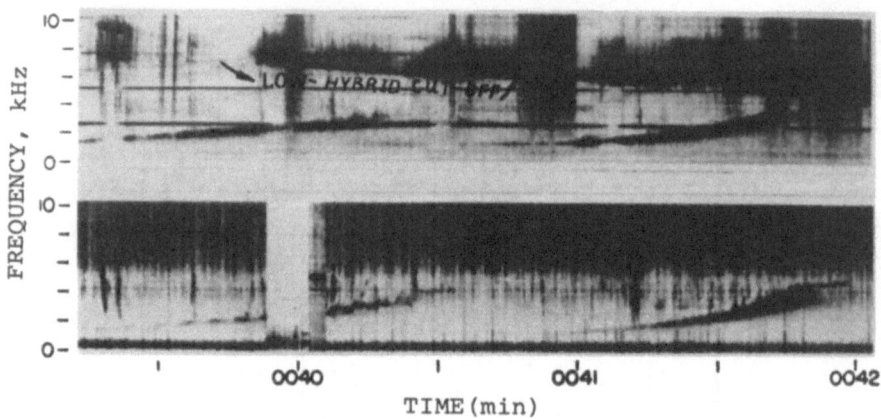

Fig. 67

should be mentioned that in the lower part of Fig. 67 there is simultaneously recorded emission of the hiss type usually observed on the Earth's surface and two emission bands of another type. Lower hybrid emission is always observed only on satellites.

Resonance effects at the lower hybrid frequency were then observed in many experiments. In the form of narrow-band emission, lower hybrid resonance waves were observed, for example, by means of the electric antenna on OGO 2 (413-1512 km) (Laaspere, Morgan, Johnson[122]). In experiments on OGO 4 (Laaspere, Taylor[123]) the ion composition was determined by means of mass spectrometers at the same time as observations were made of VLF waves. This made it possible to compare the frequency of the lower hybrid resonance that was obtained from the VLF wave spectra with data of direct measurements of the plasma parameters (see formula (1.42)). Besides cases of good agreement of the determinations of ω_L by the different methods, a discrepancy between these data was obtained in some cases. The authors assumed that this was explained by the fact that the satellite did not always record lower hybrid waves excited directly in the neighborhood of the satellite.

At altitudes higher than those in the experiments described above, narrow-band emission at the lower hybrid resonance frequency was observed on the satellite OGO 5 (perigee 291 km, apogee 147000 km, which is equal to $23R_0$) (Scarf, Fredericks, Smith, Frandsen, Serbu[124]). Figure 68 illustrates a case when these waves were observed at a distance from the center of the Earth of $R \approx 2.55R_0$ (altitude $z \sim 16 \cdot 10^3$ km). The authors emphasized that these oscillations were observed only on the electric antenna; they were not recorded on the magnetic antenna. The magnetic field H_0 and the ion density N^+ were measured simultaneously on OGO 5. It was therefore possible to compare the density N^+ obtained from the value of ω_L by means of formula (1.34) (the plasma consisted of electrons and protons) with the directly measured value of N^+. For the case shown in Fig. 68 the values $N \approx 59-76$ cm^{-3} (with maximal spread $37-88$ cm^{-3}) were obtained from the value of ω_L while tha values $N^+ \approx 66-88$ cm^{-3} were obtained from the direct measurements. One can see that there is very good agreement between these data. It should be mentioned that emissions at lower hybrid resonance frequencies were also observed on Injun 5 and OV3-3.

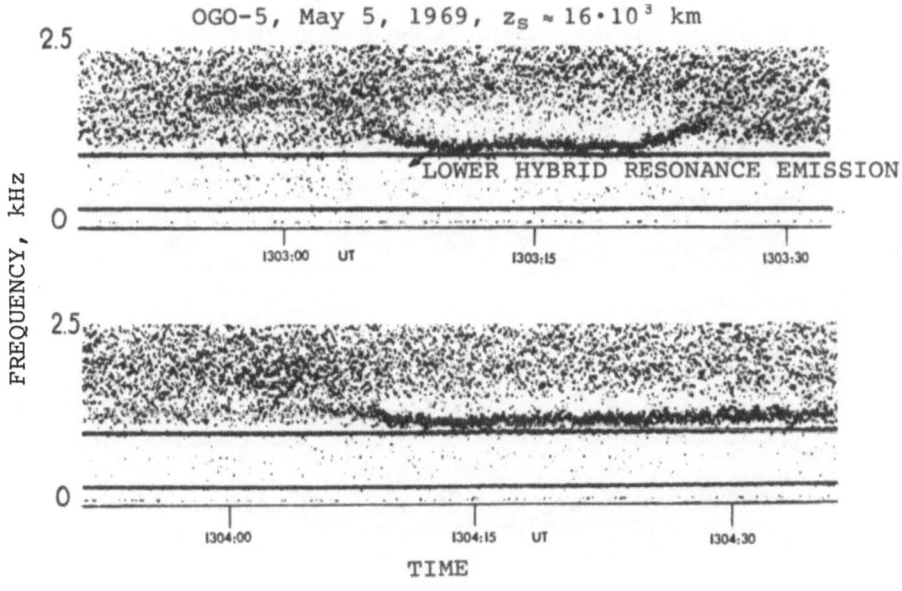

Fig. 68

4. VLF Hiss. Saucer-Shaped Emission. In the topside ionosphere, as also on the Earth's surface, one
detects different forms of broad-band emission of the
hiss type (white noise, VLF hiss) in the range of frequencies from a few to tens and more kilohertz. In a
number of cases, the lower hybrid frequency ω_L at the
point of observation lies in the frequency range of
this emission. Since one generally does not know where
these waves are excited, one does not know to what resonance branch they should be attributed; the possibility still cannot be excluded that frequently the
emission belongs to the resonance branch of LF waves
$\omega > \omega_L$ (see Fig. 4).

The most systematic study of VLF hiss has been
made in refs. 102, 103, and 104, quoted above, and in
ref. 125 (Gurnett, Frank) and in ref. 125 (Mosier, Gurnett). The main types of wave spectra observed on
Injun 5 are shown in Figs. 69 and 70. In these experiments observations were made on the magnetic and the
electric antenna and the direction from which the emission arrived was determined. In the polar zone, the
wave spectra shown in Fig. 69a were most frequently ob-

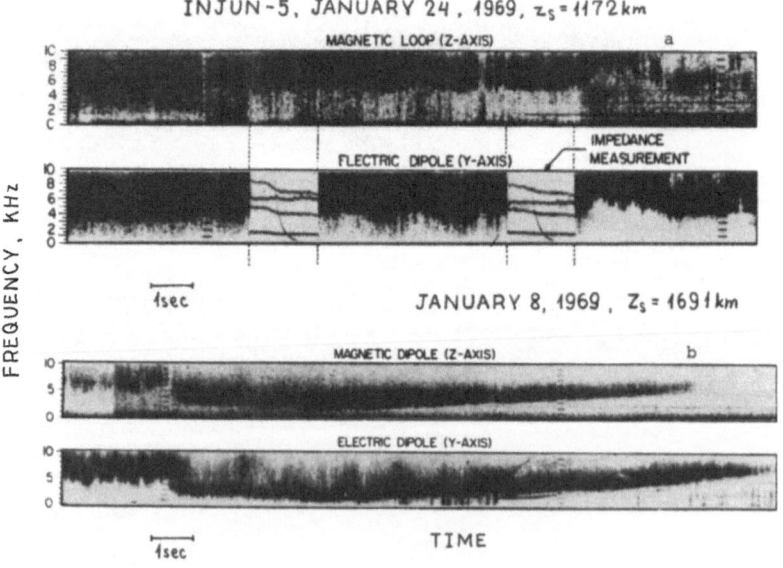

Fig. 69

INJUN-5, May 8, 1969, $z_s = 2368$ km

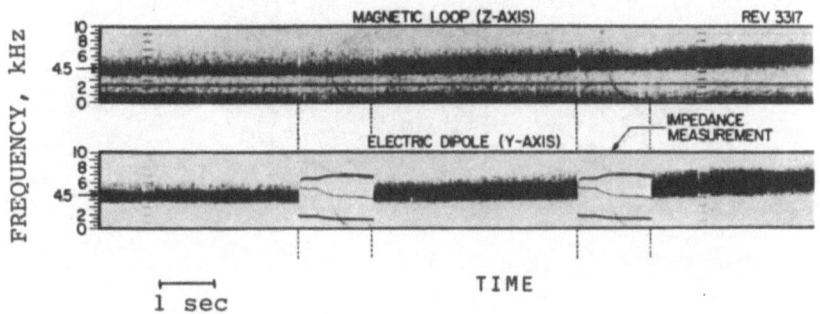

Fig. 70

served. They are characterized by large variations in
time of their frequency range; this emission is cut
off at low frequencies ω_c, the value of ω_c decreasing
with magnetic latitude and having a minimum at the latitude ~70° (see Fig. 69b). In Fig. 70 we show different type -- narrow-band VLF hiss, observed at medium
latitudes, outside of the zone of the aurora polaris.

These waves arrive at the point of observation from
both above and below. Undoubtedly, some if not all of
them are generated above the satellite, since they have
frequencies exceeding the lower hybrid frequency near
the satellite and cannot be reflected above the satellite. On the other hand, the emission coming from below
could have been reflected. However, the possibility
cannot be excluded that some of these waves are excited
below the satellite. At the present time nothing has
been published in the literature giving definite data
concerning the position of the source of this emission.
In this connection it is of interest to show here, in
Fig. 71, a color sonogram of saucer-shaped emission detected by Mosier and Gurnett;[102] this always arrived at
the satellite from below, being generated below the
point of observation. The symmetric form, of saucer
shape, of this wave packet was explained in ref. 102,
and is an effect associated with the properties of propagation in the frequency range in which this signal is
observed. Some of the wave packet is cut off outside
the surface of a cone whose axis lies along the magnetic

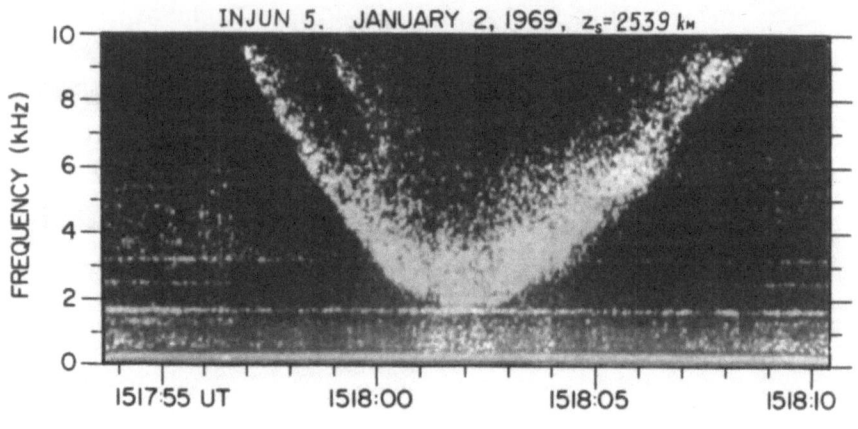

Fig. 71

field vector H_0, while the minimal frequency of the signal is equal to the lower hybrid frequency at the source.

5. Capture of Waves in the Ionosphere and Magnetosphere (Subprotonospheric, Magnetospheric-Reflected (MR) and ν-Whistlers). It was found on the Alouette satellites that short fractional-hop whistlers, which reach the point of observation directly from the surface of the Earth, are followed in the ionosphere by a second such signal that has passed through the ionosphere three times -- having been reflected once in the ionosphere above the satellite and once at the base of the ionosphere (Barrington, Belrose[127]). Detailed investigations of this effect were first made by means of the Aerobee rockets at altitudes 100-200 km and on the Alouette satellites at $z \sim 1000$ km (Carpenter, Dunkel, Walkup[128]). The results of this work showed that the later signals propagate round the ionosphere, being reflected from the base of the ionosphere at an altitude $z \sim 100$ km and above the satellite below the region of altitudes at which the protons become the principal ion constituent. Thus, it was shown that this phenomenon can be observed only on satellites, since the wave packet is trapped in the ionosphere below the protonosphere. These signals were therefore called subprotonospheric

whistlers. A series of successive short whistlers
trapped in the protonosphere, observed on Injun 5 at
the altitude 724 km, is shown in Fig. 72. On the color
sonogram, which is taken from ref. 104, one can clearly
see red (going upward) and green (going downward) signals. In the same figure one can see, behind this group
of signals, a diffuse whistler, which consists of a mixture of components going upward and downward. This is a
characteristic property of whistlers, which are strongly
scattered on inhomogeneities in the ionosphere, which
results in a multitude of propagation paths of the wave
packet from the source to the satellite.

   It is possible for VLF waves to be trapped in the
ionosphere because of the influence of the ions on their
propagation, especially in the range of frequencies
lower than the lower hybrid frequency ω_L, at which the
waves can propagate at any angle to the magnetic field
(Hines;[129] Smith;[130] Kimura;[131] Thorne, Kennel[132]).
This general theoretical interpretation, given in refs.
130-132, covers more phenomena than just subprotonospheric whistlers; in general, it also explains the "transverse" whistlers described below (Carpenter, Dunkel;[133]
Kimura, Smith, Brice[134]), magnetospheric-reflected whistlers (MR whistlers) and their modification, the so-
called ν-whistlers (Smith, Angerami[135]).

   As an illustration, in Fig. 73, we show the expect-

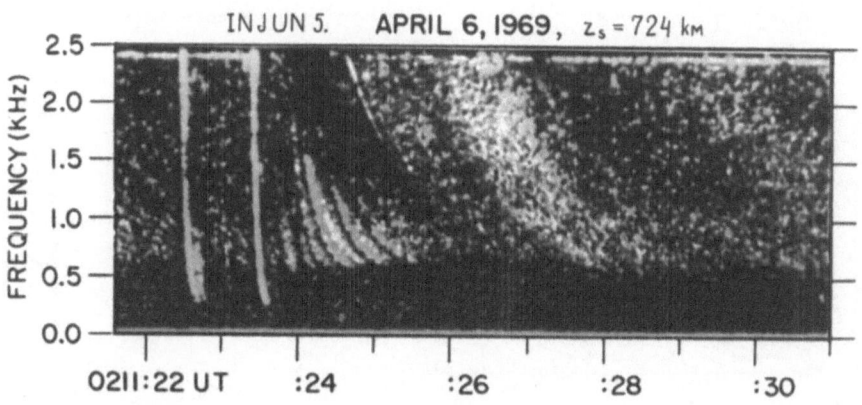

Fig. 72

ed paths of waves at the frequency 1 kHz, calculated
in ref. 134 for two models of the ionosphere under the
assumption that the initial position of a ray (the vector k_0) is vertical at the geomagnetic latitude 30°.
One can see from Fig. 73 the complicated propagation
paths of VLF waves that are possible in the near-Earth
plasma. However, one should not assume, first, that
for the complete explanation of these phenomena it is
sufficient to take into account the influence of only
the ions, nor, second, that at the present time all
features of these phenomena are understood. Rather,
the brief description given below shows that to interpret them one must take into account a number of other
factors that influence the propagation of waves in the
near-Earth plasma and that many details of these effects still remain obscure.

Smith[130] assumes that the large angles between the
wave vector and the magnetic field required for the reflection of a wave at the base of the protonosphere are
formed as a result of the influence of horizontal gra-
dients of the electron density -- they must be sufficiently large. In addition, it is assumed that the transverse
refractive index decreases with altitude.

We should here point out the main details of the
behavior of protonospheric whistlers, which can be seen,
in particular, in Fig. 72, namely: the gradual decrease

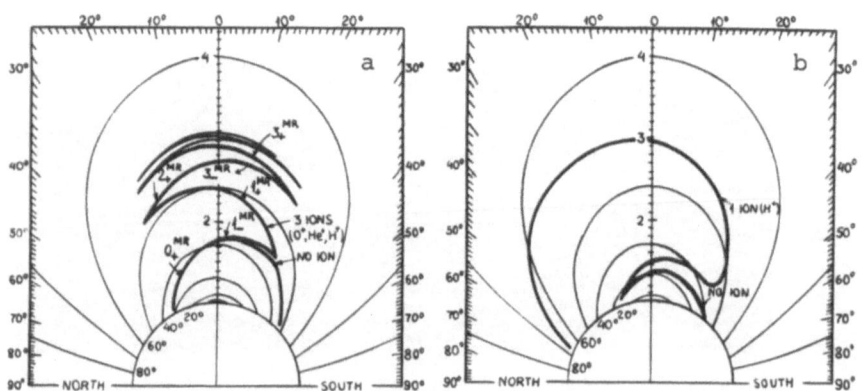

Fig. 73

in the altitude of the point of reflection of the signals above the satellite, and the reduced time delay
between the direct -- red -- and reflected -- green --
signals. In addition, the upper and lower cutoff frequencies of these signals change systematically.

The influence of transverse propagation of the
waves (relative to the vector of the Earth's magnetic
field) also explains the so-called transverse whistlers,
which are observed only on satellites, and were first
detected on the Alouette satellites.[133] This phenomenon
is as follows: at frequencies $f \approx 1\text{-}8$ kHz, which are
less than or of order of the lower hybrid frequency,
whistlers are observed in the ionosphere with a time
dependence of the frequency that is shallower than that
of the whistlers observed on the surface of the Earth.
This additional time delay varies, for example, in the
experiments of ref. 133 in the range $\Delta t \approx 0\text{-}0.22$ sec as
the magnetic latitude increases from $30°$ to $44°$. In
ref. 134 it was shown that this effect can be explained
if one assumes that a whistler follows different routes
on parts of its path, being propagated partly at right
angles to the magnetic field. However, quantitative
agreement with the experimental results is possible if
the transverse propagation of the wave occurs at least
at distances of several hundred kilometers. To obtain
an appropriate model of the ionosphere, Kimura, Smith
and Brice[134] used an attractive condition of purely
transverse propagation obtained by Hoffman,[136] namely,
the formula

$$n_{\perp q}^2 = A \cos^6 \zeta \, \omega_H^4 F(\cos \zeta /\sqrt{R}) , \qquad (3.11)$$

and the group refractive index of transverse propagation

$$n_{\perp g}^2 = (\omega_0^2/\omega_H^2) M_{eff}. \qquad (3.12)$$

Formula (3.12) follows from the transverse refractive
index (see (1.28), (1.29), and (1.42)

$$n_{\perp}^2 = [\omega_0^2/\omega_H^2 (1 - \omega^2/\omega_L^2)] M_{eff}. \qquad (3.13)$$

In (3.11), ζ is the geomagnetic latitude (it is assumed in ref. 136 that the Earth's magnetic field is a
dipole field); R is the radial distance from the center
of the Earth; A is a constant; and F is an arbitrary
smooth function.

From (3.11) and (3.13) one finds that the distri-

bution of the electron density in the ionosphere must
be described by the formula

$$N = (Am/4\pi e^2 M_{eff}) \cos^6 \zeta \, \omega_H^4 F(\cos \zeta/\sqrt{R}) . \qquad (3.14)$$

Estimates made in ref. 134 showed that the real conditions in the ionosphere could give quantitative agreement with the experimental values of Δt.

Magnetospheric-reflected whistlers, which were first
observed on OGO 1 (perigee 280 km, apogee 149400 km)
have, as we have already mentioned, a similar nature to
the subprotonospheric whistlers (see Fig. 74, ref. 135),
though they are trapped in other more extended regions
of the topside ionosphere. We should mention here that
in the literature the region of the ionosphere at altitudes $z \sim 1500-2000$ km is frequently called the magnetosphere. In this book, we stick to the terminology given
at the start of Chapter I, namely, we assume that the
upper boundary of the outer ionosphere lies at the base
of the plasmapause; in a number of cases this is at
an altitude $R \approx 3-3.5R_0$, $z \sim 20 \cdot 10^3$ km (Al'pert[14]).

The propagation of MR whistlers in the outer ionosphere is well described by the paths shown in Fig. 73
(see refs. 131 and 135). In Fig. 75 we show schemat-

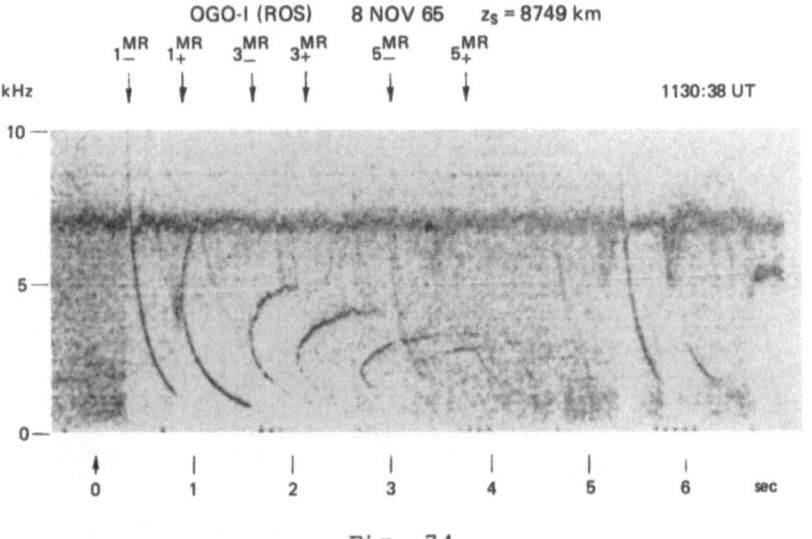

Fig. 74

ically how the wave path depicted in Fig. 73 can intersect the satellite (designated by s) a discrete number
of times and how one can observe several signals in opposite hemispheres of the Earth with respect to the source
(lightning discharge).[135] In the figures we have
used the same notation for the signals and sections of
a path. This notation changes if the satellite is in
the southern or northern hemisphere (see Figs. 74 and
75). Of course, a complete explanation of the behavior
of the experimentally observed MR whistlers is a very
complicated problem, since for each point of observation the start of the ray depends on the frequency, the
number of hops and other unknown quantities.

An interesting variant of MR whistlers, which was
also discovered for the first time in ref. 135, is
shown on the sonogram taken from ref. 135 and reproduced

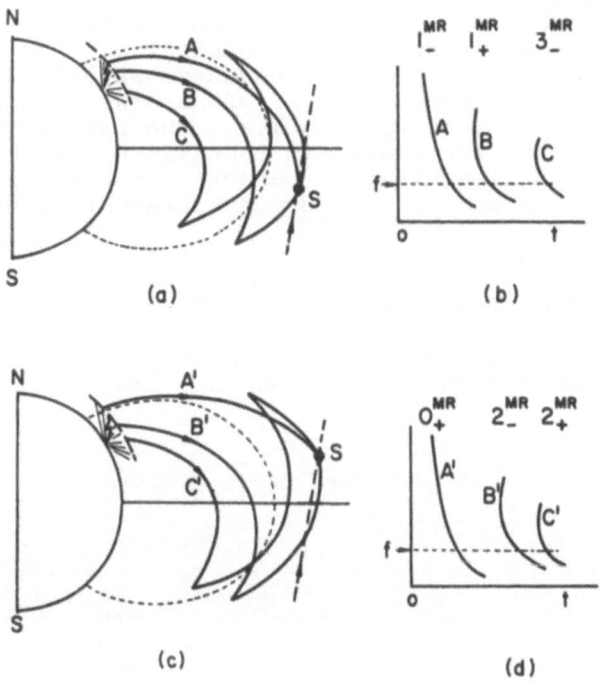

Fig. 75

in Fig. 76.   The sonogram exhibits two such whistlers
at altitude z ~ 10000 km.   These signals, which Smith and
Angerami[135] call ν whistlers (their form resembles the
Greek letter ν) are distinguished by the following features.   They are cut off at some minimal frequency ω_c.
Waves whose frequency satisfy $\omega < \omega_c$ are evidently reflected above the satellite.   On the other hand, a ν
whistler has two branches, since waves of the frequency
$\omega > \omega_c$ are reflected in the ionosphere below the point
of observation (the position of the satellite).   This
last possibility can occur if the lower hybrid frequency
ω_L near the satellite is higher than the frequency ω_c
at which both signals are cut off.   We should point out
that Thorne's study[137] of the factors responsible for
the cutoff of the ν whistlers led to interesting conclu-
sions concerning the properties of the energy spectra of
electrons in the region of energies $\varepsilon \sim 10$ keV.

     In ref. 135, as also in the subsequent experiments
on OGO 5 (see ref. 124) there were observed, in addition to the MR and ν whistlers, other modifications of
signals that do not propagate in the plasma along the
magnetic field of the Earth (nonducted waves).   Observations were also made of waves of natural origin that
are not ducted (along H_0)[124], i.e., the capture of natural emission of the plasma in bounded regions of the
plasma.   It is clear that we cannot here consider in
more detail the whole of this very interesting group of
phenomena.

§16. Results of Investigations of LF Waves $(\omega_L < \omega \lesssim \omega_H)$

     In the foregoing section we have already described
waves that belong to the LF frequency range.   Namely,

OGO-I, OCTOBER 17, 1964, $z_s \approx 10000$ km

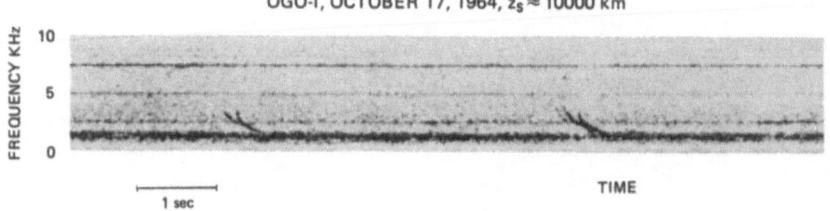

Fig. 76

waves of frequency $\omega > \omega_L$ up to $\omega \sim \Omega_0$, to which the resonance branch of waves excited in a cold plasma (see Fig. 4) belongs, and also the branch of fast ion-acoustic waves (Figs. 5 and 6) and the branch of slow electron-acoustic waves in a nonisothermal plasma (see §4.2) which also belongs in the LF frequency range. In this section we shall consider predominantly wave processes ($\omega \gg \omega_L$) whose behavior (excitation, propagation, damping) is due principally to oscillations of electrons in the plasma.

1. Electron Whistlers. The most popular representatives of these waves are long electron whistlers observed in regions on the surface of the Earth or in the ionosphere that are magnetically conjugate to the source. An example of a sonogram of such an electron whistler, obtained in the satellite Vanguard 3, is shown in Fig. 77. It was noted in experiments on Injun 3 that at certain periods 20 whistlers per minute were detected by the satellite. [107] During the same period of observation, instruments of the same sensitivity on the Earth's surface detected only three whistlers in 15 hours!

As we have already mentioned above, in this book data on whistlers are considered only to illustrate some of the effects observed in the near-Earth plasma (see refs. 89 and 90). However, it is expedient to list here the following fundamental and important facts discovered in their investigation.

1. It was whistlers that first made possible the discovery and most detailed investigation of the so-called knee at the upper boundary of the ionosphere, namely, the region of the plasmapause, where the electron densi-

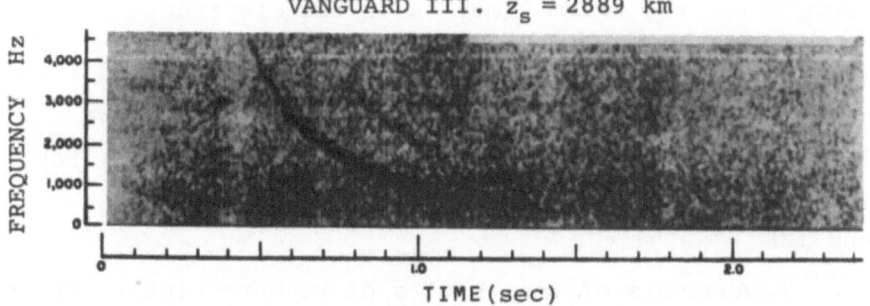

VANGUARD III. $z_s = 2889$ km

Fig. 77

ty falls in a number of cases by a factor of approximately 10-100 in a range of distances $\Delta R \sim 0.15 R_0$ at a
distance of $R \sim (3.3-5) R_0$ from the center of the Earth
in the equatorial plane (Carpenter;[138], [139] Angerami,
Carpenter[140]). It was shown subsequently that the
boundary of the plasmapause and the drop in the electron
density vary in even wider limits (see, for example,
ref. 141, Carpenter, Park, Taylor, Brinton).

2. In recent years it has been established difin-
itively that the propagation of whistlers along H_0 is
determined by the influence of longitudinal inhomogeneous formations that arise in the near-Earth plasma
along the lines of force of the geomagnetic field H_0. As
a consequence of this, electron whistlers are preferen-
tially cut off at the frequency $\omega_c = \omega_{H0}/2$, where ω_{H0} is
the proton gyrofrequency of the electrons in the equa-
torial plane at the apogee of the signal path (Carpenter[142]). This effect was predicted theoretically a long
time ago in papers in which a study was made of the
trapping of waves in inhomogeneities formed along H_0
(Smith, Helliwell, Yabroff;[143] Smith[144]). Until recent
years it was assumed that whistlers are cut off primarily
at the gyroresonance frequency ω_H of the electrons and
that the cutoff is described by the spatial coefficient
of cyclotron damping of the electron wave:

$$\kappa_{He} = \tfrac{1}{2}(\sqrt{\pi} c/v_e)[(\omega_H - \omega)/\omega] \exp(-z_{He}^2), \qquad (3.15)$$

where

$$z_{He} = [(\omega_H - \omega)/\omega] \cdot (c/v_e n_{20}) \gg 1. \qquad (3.16)$$

In a number of studies of whistlers at their path apogee,
a kinetic correction has been used in the refractive index describing a whistler-mode wave. Namely, with allowance for the thermal motion of the electrons

$$n^2 = [\omega_0^2/\omega(\omega_H \cos\theta - \omega)](1 + \delta), \qquad (3.17)$$

where if $\theta = 0$ and the condition (3.16) is satisfied

$$\delta = \tfrac{1}{2}(v_e^2/c^2)[\omega_0^2 \omega/(\omega_H - \omega)^3] \qquad (3.18)$$

(for more detailed formulas for κ_{He} and n see refs. 3
and 12).

3. Analysis of the results of recent whistler observations on OGO 3 (Angerami[145]) has shown that under

the conditions of these experiments:

a) the width of the longitudinal inhomognenous formations varies in the range $(0.035-1.070)R_0 \approx 200-400$ km;

b) the inhomogeneous formations are separated by distances of $(0.017-0.18)R_0 \approx 100-1000$ km;

c) the length of the inhomogeneous formations at the equator is of order $0.3R_0 \sim 2000$ km;

d) the inhomogeneities are areas of accumulation and not depletion of the electron density ($\delta N = N - N_0 > 0$, where N_0 is the unperturbed electron density.

4. Observations of whistlers on the Earth's surface revealed on the sonograms of the whistlers bands of waves excited in the near-Earth plasma by the whistlers themselves; these are known as whistler-triggered emissions (Heliwell[177]). Study of these narrow-band packets showed that they are observed predominantly in two frequency ranges and have the following properties (Helliwell,[89] Carpenter[142]).

A. One of the wave modes is excited in the neighborhood of the frequency f_c at which the whistler is cut off, namely, for $f \lesssim f_c \approx f_{H0}/2$ where f_{H0} is the gyrofrequency at the apogee of the magnetic line of force along which the whistler propagates. The frequency of this wave packet increases very slowly with the time; it has a width of order 50 Hz. In a number of cases the duration of the emission reaches 20 sec.

B. The other type of narrow-band signal is excited near the tail of the whistlers at the frequency $f \approx f_{H0}/6$! The value of f of this wave packet increases rapidly and appreciably with the time (rising tone) and also has a width of order 50 Hz. Sometimes these signals have a complicated form (hooks), i.e., also a branch with decreasing frequency.

Recent experiments have also revealed a narrow-band emission of the near-Earth plasma stimulated by terrestrial long-wave radio stations (Helliwell, Katsufrakis, Trimpi, Brice,[178] Helliwell, Katsufrakis, Kimura[179]); these are known as artificially stimulated emissions. These emissions are excited in the plasma sometime after the arrival of the radio waves in the plasma, which is revealed from the traces of dashes in Morse code signals. The more powerful is the radio transmitter, the earlier these waves are excited. If the dashes are sufficiently long, excitation is observed even for radio transmitters of very low power (Kimura[180]). These wave packets are more readily excited when the frequency of the radio transmitter satisfies f ≈ $f_{H0}/2$. The frequency at which the excitation occurs is evidently somewhat higher than the frequency f of the radio transmitter. The frequency of the wave packet is then strongly changed in different ways. In some cases it increases with the time (rising tone) and in some cases it decreases (falling tone). Cases are also observed when the wave packet frequency first decreases and then increases or vice versa (branching spectrum).

Different theoretical explanations have been put forward for this narrow-band emission of the near-Earth plasma stimulated by external sources of electromagnetic fields (see, for example, Helliwell,[181] Kimura,[182] Matsumoto[183]).

2. LF Chorus, Hiss, and Other Emissions Generated near the Geomagnetic Equator. In recent years in a series of experiments on the satellites OGO 1 and OGO 3 investigations have been made of different types of LF emission whose frequencies are less than the electron gyrofrequency ω_H in the neighborhood of the satellite (Burtis, Helliwell;[146] Dunckel, Helliwell;[147] Dunckel, Ficklin, Rorden, Helliwell[148]). Measurements were made on these satellites with only a magnetic antenna.

It was established that these emissions are generated primarily in the neighborhood of the geomagnetic equator at the frequency $\omega \lesssim \omega_{H0}/2$ (ω_{H0} is the gyrofrequency at the equator) in the transition region between the plasmapause and the magnetosphere at values L ≈ (4-10)R_0. The intensity of the emissions and the probabilities of their being generated decrease with increasing distance from the Earth. When the satellite passed through the front of the shock wave formed around the Earth in the magnetosphere, no single case of detection of LF emission was made in these experiments.

Examples of sonograms of emission of the hiss and chorus type, obtained on OGO 1 and OGO 3, are shown in Figs. 78 and 79. Hiss emissions were observed in the range from 0.3 kHz (lower frequency of the receiver) to 3 kHz. Usually, the emission was cut off abruptly at the higher frequency. The nature of the chorus emission (Fig. 79) changed with the distance from the Earth -- the emission bands becoming more sporadic. Thus, in Fig. 79 (in which we have also given the dependence of

OGO-I, $z_s \approx 10^4$ km

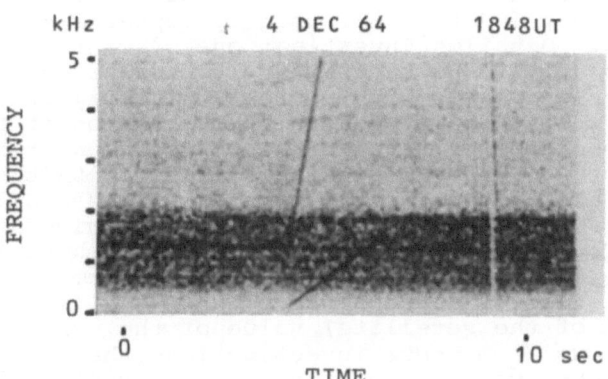

Fig. 78

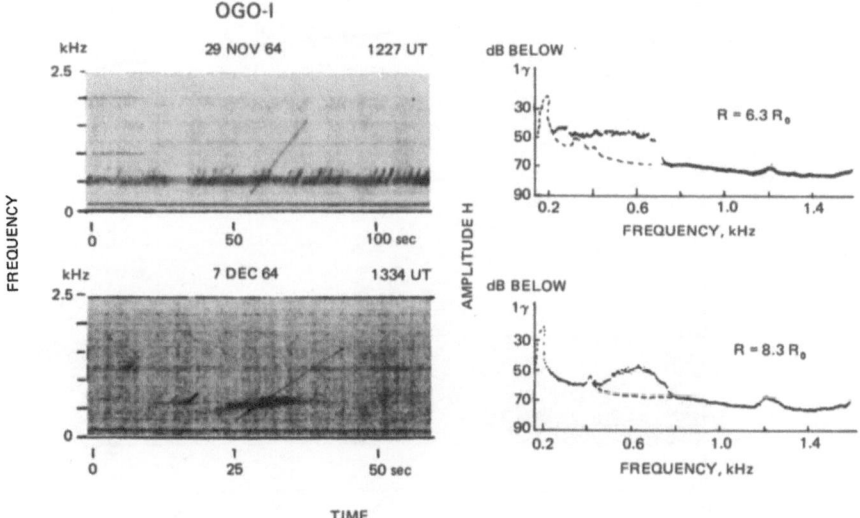

Fig. 79

the field amplitude H on the frequency) it can be seen
that at a distance $R \approx 52 \cdot 10^3$ km from the center of the
Earth only one emission band was observed in the interval $\Delta t \sim 50$ sec. The duration of such emission bands is
$\Delta t \sim 10$ sec, and they are usually repeated after several
minutes. Detailed investigations of chorus emissions
(see ref. 146) showed that the central frequency of this
emission decreases with decreasing geomagnetic latitude
and is well correlated with the variation of the minimal
gyroresonance frequency of electrons along the line of
force of the geomagnetic field, i.e., with the value of
ω_{H0} at its apogee (on the equator). The results of a
comparison of $f_{H0} = \omega_{H0}/2\pi$ with the emission detected on
OGO 1 (sometimes a mixture of hiss and chorus were observed) are shown in Fig. 80. In the same figure we
have plotted the dependence of the local (in the neighborhood of the satellite) value of the gyrofrequency.
It follows from these investigations that the frequency
of the observed chorus emissions generally varies in the
limits $\omega \approx (0.2-0.5)\omega_{H0}$. However, analysis of the exper-

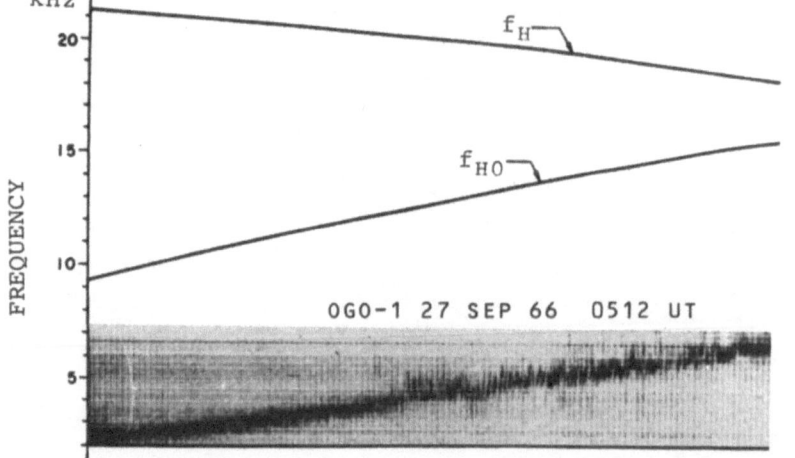

Fig. 80

imental data and the calculations made in ref. 146 led
to the conclusion that the emission is actually gener-
ated in the equatorial plane at the frequency $\omega \sim \omega_{HO}/2$,
but the waves propagate to the point of observation along
paths that deviate from a line of force of the magnetic
field. The actual paths of these waves are more elongated than the lines of force. That such paths could
exist for whistler-mode waves had already been established theoretically (see, for example, ref. 132). Therefore, if they are associated with lines of force of the
geomagnetic field in the interpretation of the experimental data, underestimated values of the ratio ω/ω_{HO}
are obtained, especially at high geomagnetic latitudes.

Similarly, it was established in these experiments
that the upper cutoff frequency of the different emissions varies with the position of the satellite and is
also proportional to ω_{HO}. This follows, for example,
from Fig. 81, in which we show a sonogram (and details
of it for three short intervals of time) taken in the
region $L \approx (6.0-8.5)R_0$. With increasing L, the value of
ω_{HO} decreased, and the upper limit of the cutoff frequency of the emission decreased proportionately.

Burtis, Helliwell, and Dunckel[146, 147] concluded
that these LF emissions were generated in the equatorial
plane, in agreement with the theory developed in ref.
149 (Kennel, Petschek).

An interesting experimental investigation of the
fine structure of different types of emission was made
recently for the first time in ref. 150 (Coroniti, Fredericks, Kennel) by means of a spectrum analyzer with a
high time resolution. The paper ref. 150 contains the
results of the analysis of sonograms of LF chorus emissions observed at a distance of $R \approx 30 \cdot 10^3$ km from the
center of the Earth near the equatorial zone (Fig. 82).
The sequence of traces in this figure follow at intervals
of time $\Delta t = 12.5$ msec, and it describes the structure of
the spectrum and its evolution in time. Coroniti, Fredericks, and Kennel conclude that the chorus spectrum
consists of narrow-band modes of widths $\Delta f \approx 20-30$ Hz and
frequency-modulated waves. There is no doubt that further investigations of the fine structure of spectra
will enable one to study more deeply the nature of the
various wave modes observed in the near-Earth plasma.

We may mention that in experiments on OGO 1 two
new types of emission were also observed -- broad-band

OGO-I

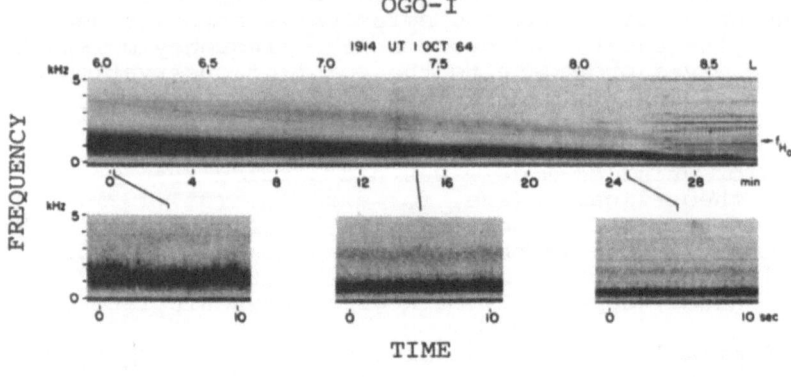

Fig. 81

and an emission that was cut off at a frequency ~20 kHz
(highpass); Kennel and Petschek[148] assume that these
emissions are generated near the satellite. We shall
not here describe these emissions, but refer the reader
to ref. 148.

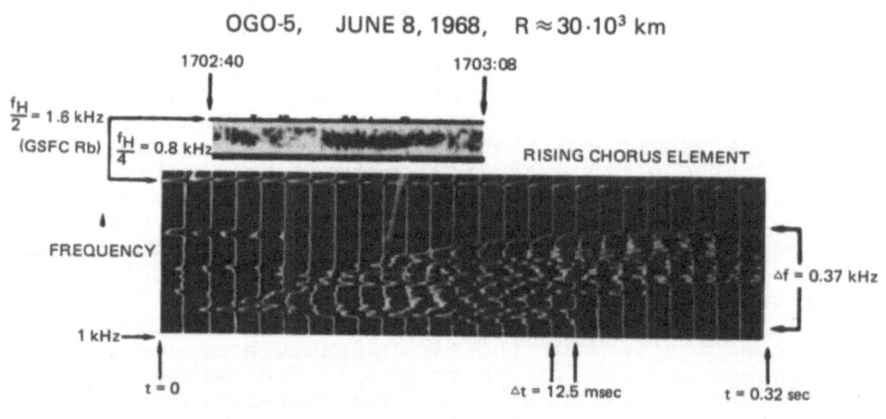

Fig. 82

§ 17. Results of Investigations of HF Waves $(\omega > \omega_H)$

Above, in §13, we have already mentioned the main
results and features of the emission of the near-Earth
plasma in the HF frequency range observed on satellites.
We shall here consider briefly some concrete examples of
these results.

The theoretically expected high-frequency resonances
(see Fig. 3 and §4) and the HF emissions that have not
yet received a clear theoretical interpretation (which
we consider here) were obtained in experiments in basic
ally three regions of the natural plasma: in the ionosphere in the altitude range z ~ 800-2000 km, between
the plasmapause and the lower boundary of the magneto-
sphere at distances $R \approx 30\text{-}50 \cdot 10^3$ km from the center of
the Earth, and in the solar wind when the satellite was
approximately 100-150 thousand to a million kilometers
from the Earth. In the experiments in the outer ionosphere, data were obtained on waves and oscillations of
the plasma that were necessarily excited under the shock
effect of radio pulses transmitted from the satellite.
In the case of the magnetosphere and the solar wind, the
emission that was observed was predominantly generated
by streams of particles emitted by the Sun.

1. HF Resonances Observed in the Topside Ionosphere.
Data on resonances in the ionosphere in the altitude
range 800-3000 km were first obtained on the Alouette
satellites (see ref. 151). These satellites carried
multifrequency pulse radio transmitters (ion probes),
which, ranging over a wide frequency interval, trans-
mitted pulses (narrow wave packets with $\Delta\omega \ll \omega$) in the
neighborhood of the carrier frequency ω and simultaneously received waves at these frequencies that were reflected or generated in the ionosphere. It was established for the first time in these experiments that when
the frequency of the transmitter changes resonance frequencies of the plasma -- the corresponding packets of
characteristic oscillations of the plasma -- are excited
in the transmitter. They appear in the altitude-frequen-
cy characteristics in the form of spikes, whose lifetime
varies from case to case and depends on the type of resonance. The corresponding characteristic, obtained on
Alouette 2, is shown in Fig. 83 (Calvert, McAfee[152]).
In the left hand part of the figure, along the ordinate,

we have plotted the scale of so-called virtual ranges, $z_V = c\Delta t/2$, where c is the velocity of electromagnetic waves in vacuum and Δt (right-hand scale) is the delay time of the reflected waves or the lifetime of the transmitted waves. On the horizontal axis along the bottom we have plotted the frequencies over which the ion probe ranged during a short interval of time. Along the top we have plotted the values of the resonance frequencies ($\omega_H = 2\pi f_H$, $\omega_0 = 2\pi f_0$, $2\omega_H$, etc) at which spikes were observed. In the case shown in Fig. 83, gyroresonances were observed up to the third harmonic (ω_H, $2\omega_H$ and $3\omega_H$) two upper hybrid resonances (ω_U and $2\omega_U$), and Langmuir waves at ω_0. In the Alouette experiments, gyroresonances $s\omega_H$ were observed from s = 1 to s = 16 (Lockwood[153]). We have already mentioned above that in some cases "half" resonances were observed: $\omega = \omega_H/2$, $\omega_0/2$, $3\omega_H/2$ (Nelms, Lockwood[154]). In one of the recent studies, Oya[155] made detailed studies of the resonances of "diffuse type" ω_{D1}, ω_{D2},... in the intermediate regions between the integral gyroresonances $s\omega_H$ and $(s + 1)\omega_H$. These were dubbed diffuse resonances by Nelms and Lockwood[154] because these signals are usually smeared out, occupying a wide frequency interval (see Fig. 84) and their central frequency does not correspond exactly to a half-integral value $(s+ 1)\omega_H/2$. Results obtained by analyzing many measurements showed that in the cases when $\omega_0 < 1.8\omega_H$ (in these experiments there were 900 such cases) resonances were not observed at the

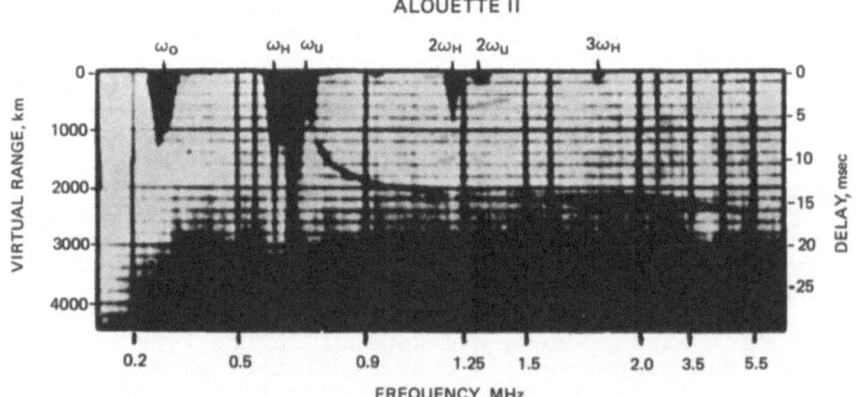

Fig. 83

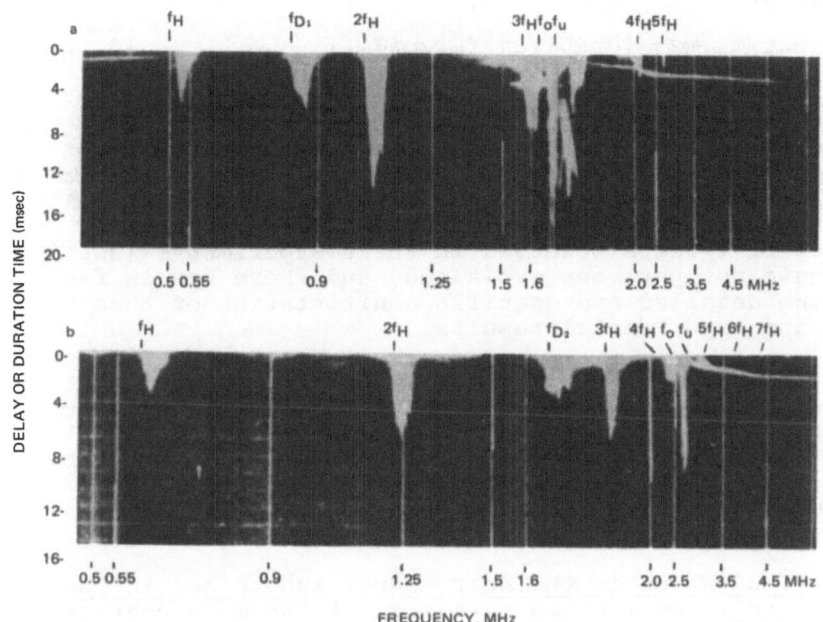

Fig. 84

frequencies ω_{Ds}. In the opposite case, when $\omega_0 > 1.8\omega_H$ they were always observed. The resonances ω_{Ds} were observed only up to $s = 5$. In some cases, the resonance region ω_{Ds} was split into two signals. Generally, the frequencies ω_{Ds} of the resonances varied in the following limits:

$$\omega_H - 2\omega_H : \omega_{D1} \approx (1.4 - 1.85)\omega_H, \quad \omega_{D1} \approx (1.7 - 1.9)\omega_H, \quad \omega_0 \approx (2.0 - 3.7)\omega_H$$

$$2\omega_H - 3\omega_H : \omega_{D2} \approx (2.4 - 2.9)\omega_H, \quad \omega_{D2} \approx (2.6 - 3.0)\omega_H, \quad \omega_0 \approx (3.4 - 4.8)\omega_H$$

$$3\omega_H - 4\omega_H : \omega_{D3} \approx (3.6 - 3.9)\omega_H, \quad\quad - \quad\quad \omega_0 \approx (4.5 - 5.8)\omega_H$$

$$4\omega_H - 5\omega_H : \omega_{D4} \approx (4.5 - 4.9)\omega_H, \quad\quad - \quad\quad \omega_0 \approx (5.4 - 6.7)\omega_H$$

$$(3.19)$$

It is assumed that the resonances ω_{Ds} are electrostatic

waves (see ref. 156; Warren, Hagg) and are described
by the following formula obtained in ref. 157 (Dougherty,
Monaghan):

$$\omega_{Ds}/\omega_H = s + (0.464/s^2)(\omega_0^2/\omega_H^2). \tag{3.20}$$

In ref. 157, as in earlier papers (see, for example, Fejer and Calvert[158] and Crawford, Harp, and
Mantei[159]) theoretical studies were made of the resonances observed on the Alouette satellites and stimulated
by high-frequency fields generated around the satellite
by the straight transmitting antennas. Hitherto, a
number of effects observed in these experiments (see §
13) have not yet been explained, and there has in fact
been no detailed and specific confrontation of theoretical and experimental results.

It should be noted that the main feature of the
resonances in the outer region of the ionosphere considered here is their short lifetime: they last for
only $(5-10)10^{-3}$ sec. In contrast, the emission in the
near-Earth plasma considered in the next subsection
was usually observed for several or even tens of minutes.

2. HF Waves in the Near Magnetosphere and in the
Solar Wind. We mention here some of the more characteristic results of experiments obtained in the HF frequency range.

The satellite OGO 5 revealed narrow-band emissions
at the frequencies $\omega \approx (s+1)\omega_H/2$ (emissions were observed
up to $\omega \approx q\omega_H/2$), most frequently at the frequency $\omega \approx
3\omega_H/2$; these emissions were usually observed for periods
of many minutes (Kennel, Scarf, Fredericks, McGehee,
Coroniti[160]). They were observed at distances from the
Earth of $R \approx (30-50)10^3$ km between the plasmapause and
the magnetosphere. Analysis of the experimental data
showed that the emissions were excited in the neighborhood of the geomagnetic equator. The values of the electric field E at the frequency $\omega \approx 3\omega_H/2$ were always
large. They varied in different cases from several
millivolts to several tens of millivolts per meter. A
sonogram of the emission at frequencies $\omega \approx 3\omega_H/2$ observed in one of the experiments at a distance of $R \approx (40-
50)10^3$ km from the center of the Earth and the depend-
ence of the field strength of these waves is shown in
Fig. 85 (Scarf, Fredericks[161]). Analysis of the fine
structure of this emission, made in ref. 150 in the

time interval $\Delta t = 0.42$ sec every 12.5 msec when $R \approx 42 \cdot$
10^3 km and at magnetic latitude $\lambda_M = -4°$ is shown in Fig.
86. The (ω,t) diagram obtained in ref. 150 for this
time interval shows that the emission at the frequency
$\omega = 3\omega_H/2$ evidently consists of several modes in a narrow frequency range $\Delta f \approx 200$ Hz. In these experiments it
was shown that the emission at the frequencies $\omega = (s + 1)$
$\omega_H/2$ consists of longitudinal ($k_0 \| E$) waves. The excitation mechanisms of these waves and their theoretical explanation have not yet been given.

Interesting results of observations of HF waves on
the satellite IMP 6 have recently been described (Shaw,
Gurnett[176]). In these experiments, beginning at distances from the Earth's center of $z \sim 3R_0$ (~20000 km), in
the region adjoining the upper boundary of the ionosphere
(plasmapause), to $z \sim (8-9)R_0$ (in the magnetosphere),
emission bands were observed in a continuous frequency
interval corresponding to the high-frequency resonance

OGO-5, AUGUST 15, 1968, $z_s \approx 48 \cdot 10^3$ km

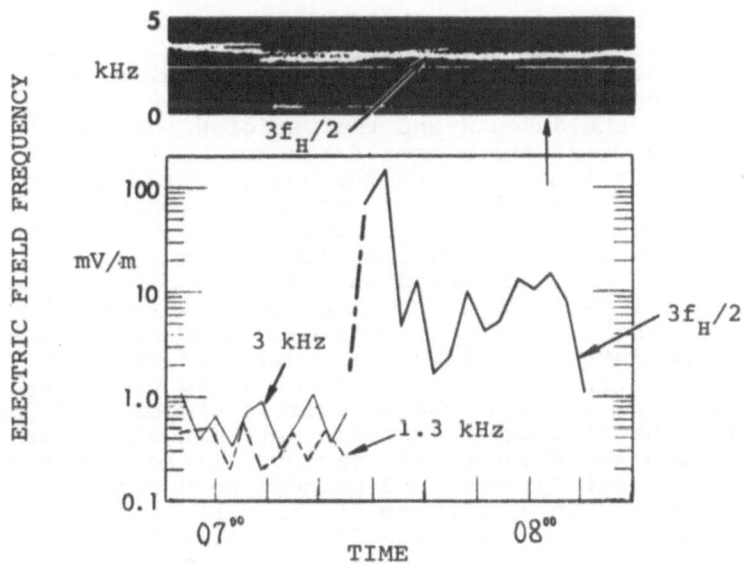

Fig. 85

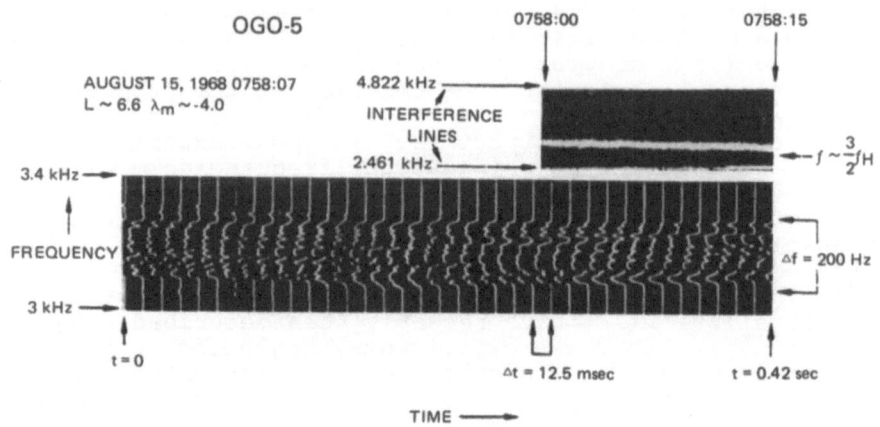

Fig. 86

branch of plasma waves $\omega_1(\theta)$ described by formulas (1.31)
and (1.32) (see Fig. 3). This emission was observed
only with the electric antenna (longitudinal waves).
In accordance with theory, it was cut off, at an angle
$\theta \approx 0$ between the antenna and the vector H_0 of the Earth's
magnetic field, at the plasma frequence ω_0 (when
ω_0 was greater than ω_H); at the angle $\theta \approx \pi/2$ the emission was cut off at the upper hybrid frequency $\omega_U =
(\omega_0^2 + \omega_H^2)^{\frac{1}{2}}$ (see Fig. 57). The field strength of these
waves did not exceed 10 μV per meter (energy density
$\sim 10^{-21}$ erg/cm^3). They could therefore be observed only
under conditions of a quiescent state of the magnetosphere, and their observation was made possible by the
use of long antennas: two horizontal ones of length
53.5 m and 92.5 m and a vertical one (directed along
the axis of rotation of the satellite) of length 7.7 m.
Since the cutoff frequencies were determined very accurately and the value of the magnetic field H_0 was also
measured accurately, Shaw and Gurnett were able to use
the results of these measurements to make an accurate
determination of the electron density of the regions of
plasma traversed by the satellite by two methods: from
the values of ω_0 and ω_U. One of their profiles of the
electron density $N(z)$ above the plasmapause in the mag-

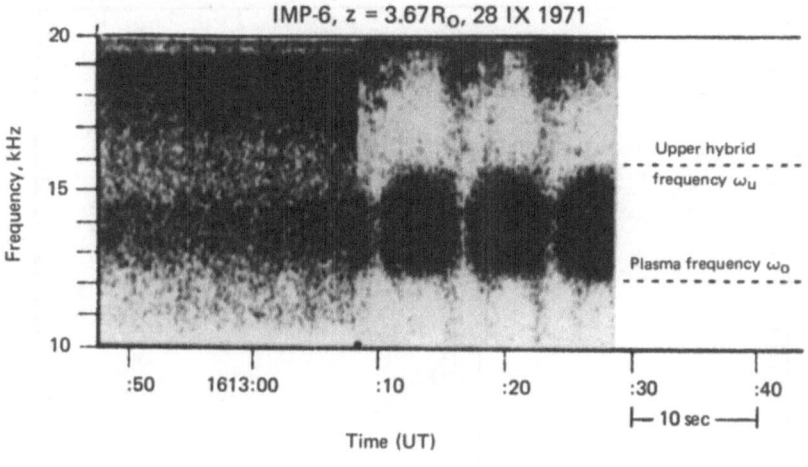

Fig. 87

netosphere is shown in Fig. 88. There is no doubt that
the accuracy of this resonance method of determining
N(z) is higher than the accuracy of other methods, since
the results of the measurements do not depend on perturbations of the plasma in the neighborhood of the satellite, whose linear dimensions are less than the wavelengths of the excited plasma waves, and the frequencies
ω_0 and ω_U can be determined very accurately. Shaw and
Gurnett suggest that incoherent Cherenkov emission of
superthermal electrons were observed in these experiments.

Resonant excitation of longitudinal Langmuir waves
in the neighborhood of the frequency ω_0 described by
the dispersion equation (1.44) has been observed in different experiments. Of interest are the corresponding
results obtained in the solar wind on the space probe
Pioneer 8 with heliocentric orbit (aphelion equal to
1.09 astronomical units and perihelion of order 1.0
astronomical units) at distances from Earth of order
10^6 km. A trace of these waves at the frequency 20 kHz

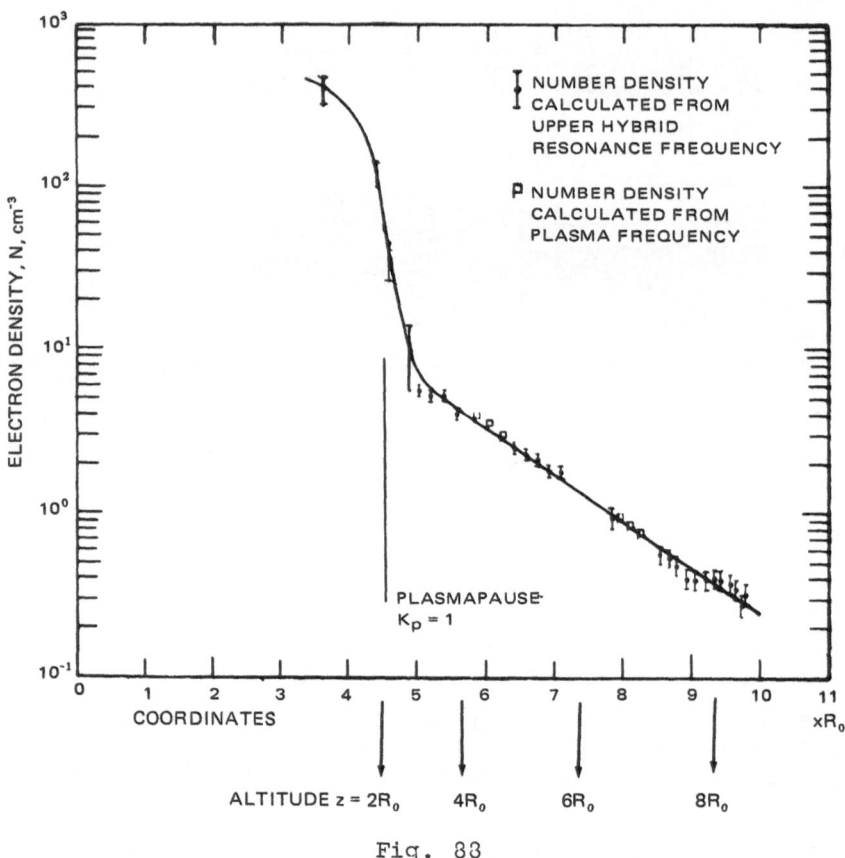

Fig. 88

is shown in Fig. 89.[117] In this experiment the wavelength
of these plasma oscillations was determined, giving values $\Lambda \approx 300$-500 m! The fact that longitudinal waves are
excited and transformed in the solar wind (see below) is
interesting, and is an indication that complicated processes take place in this region of the plasma. According to estimates, the energy density of these waves
was of order 10^{-14} -10^{-15} erg/cm^3. At the same time,
the energy density of the particle streams was $N(MV_0^2)$ ~
10^{-9} erg/cm^3 and $N\kappa T_e \approx 10^{-10}$ erg/cm^3. Thus, the inten-

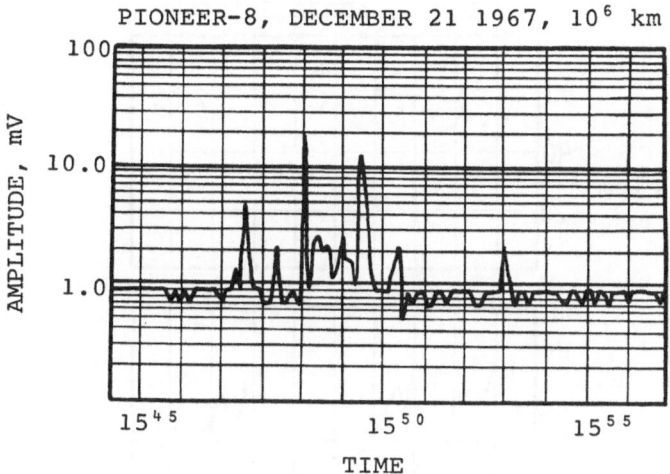

Fig. 89

sity of the Langmuir waves was only a small fraction of
of this.

    In the experiments on OGO 5 in the interplanetary
medium at distances from the Earth of order 100-150
thousand kilometers (near the apogee of the satellite)
transverse electromagnetic waves at the frequency f =
70 kHz were discovered in the solar wind. This frequency
lay in the interval $\omega_0 < \omega < \omega_U$, i.e., it was near the
electron plasma frequency (Scarf, Fredericks, Green,
Neugebauer[162]). It should be noted that these observations were made during the period of a strong flare,
when the solar wind density N exceeded its normal value
by a factor 8-10. Usually, the plasma frequency of the
solar wind is 14-30 kHz. The electric and magnetic components of the field were measured simultaneously on
OGO 5 and probe measurements were also made of the plasma
density N. Figure 90 shows the results of measurements
of E, H, and N obtained in one experiment. In the same
figure we have plotted the values of the electron density N(E,H) calculated from the relation n = cE/H, which
relates the electric and magnetic components of trans-

OGO-5, APRIL 5 1968

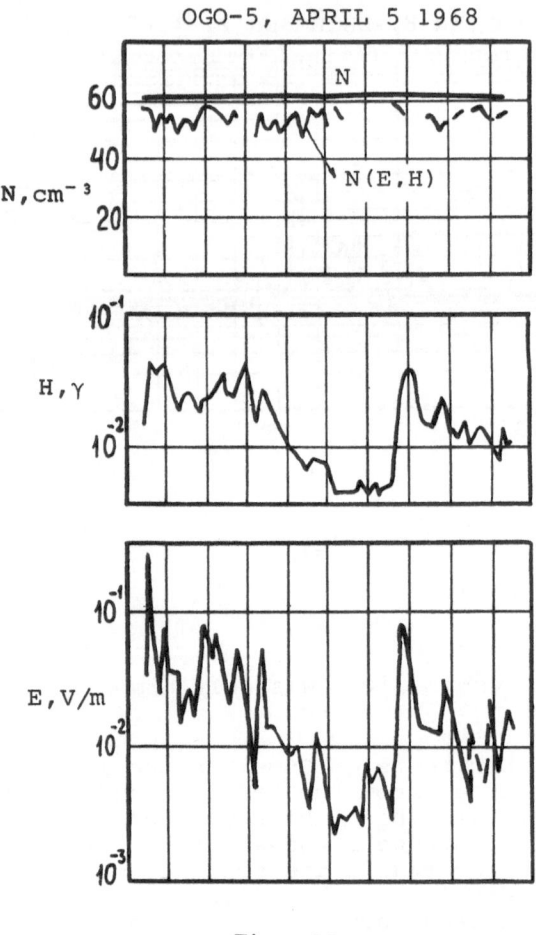

Fig. 90

verse electromagnetic waves. It can be seen that there
is very good agreement between the values. In the recent
paper ref. 125 similar data were given for a determination
of N(E,H) during a strong storm at the frequency f = 70
kHz, at which transverse electromagnetic waves were observed on OGO 5 at a distance from the center of the

Earth of $R \approx 4.7 \cdot 10^3$ km. The values obtained were $N \approx 63$ cm^{-3} and $N(E,H) \approx 61$ cm^{-3}.

Thus, the data given above indicate that the electromagnetic waves (transverse, $\mathbf{k}_0 \perp \mathbf{E}$, H_0) trapped in the solar wind are evidently the result of transformation of longitudinal resonant Langmuir waves excited in the solar wind. The theoretical explanation of the actual mechanisms of excitation of these waves and their transformation will undoubtedly be the subject of further investigations, alongside the detailed theoretical analysis of similar very interesting experimental data.

§18. Energy Densities of Different Types of Waves

It is evident from the data considered in the foregoing sections that investigations of wave processes in the near-Earth and interplanetary plasma have been carried on extensively in recent years and that the information currently available on these phenomena in the literature is very extensive. At the same time, the frequent absence of the results of simultaneous measurements of, for example, the densities of charged particles and other parameters of the plasma prevents one determining the energy fluxes W_E and W_H of these waves, which depend on the refractive index of the plasma, namely, the values of

$$W_E = 2.66 \cdot 10^{-3} n E^2 \, (V/m) \quad W/m^2 \tag{3.21}$$

and

$$W_H = (2.4 \cdot 10^{-4}/n) H^2 \, (\gamma) \quad W/m^2, \tag{3.22}$$

where the magnetic field is expressed in gammas. Generally, under different conditions and at different frequencies the electric and magnetic field strengths vary in the following limits: $E \sim 10^{-4}-1$ V/m and $H \sim 10^{-3}-1$ γ. However, cases are known in which E reached tens of volts per meter and H several gammas. For reference, we mention here that generally $W_E \approx 10^{-8}-10^{-11}$ W.m^{-2}.Hz^{-1}.

It is easy to extract data from the energy density from the results of different experiments:

$$W_E = 4.4 \cdot 10^{-11} E^2 \, (V/m) \quad erg/cm^3,$$
$$W_H = 4 \cdot 10^{-16} H^2 \, (\gamma) \quad erg/cm^3. \tag{3.23}$$

We give here brief data on the W_E and W_H values observed in different experiments.

In the range of ELF waves, the energy density of transverse waves (subscript $\perp$) at frequencies $500 > f > 10$ Hz usually varies in the range

$$W_\perp \sim 10^{-15} - 10^{-18} \ \text{erg/cm}^3. \qquad (3.24)$$

The value of $W_\perp$ increases strongly with decreasing frequency and for $f < 10$ Hz

$$W_\perp \sim 10^{-11} - 10^{-15} \ \text{erg/cm}^3. \qquad (3.25)$$

It should be pointed that in the transition zone between the magnetosphere and the interplanetary medium (magnetosheath) values were obtained for ELF waves at frequencies $f \lesssim 1$ Hz $\lesssim \Omega_H/2\pi$ for the energy density of transverse waves that satisfied $W_\perp \gtrsim 10^{-15}$ erg.cm^{-3}. Hz^{-1} (Smith, Holzer, McLeod, Russell[185]). In the interplanetary medium for ELF waves at $f \approx 3 \cdot 10^{-4} - 0.5$ sec$^{-1} \lesssim \Omega_H/2\pi$ in other experiments the energy density decreased with increasing frequency approximately in proportion to $f^{-3/2}$ and varied from $W_\perp \approx 2 \cdot 10^{-11}$ erg.cm^{-3}.Hz^{-1} to $W_\perp \approx 4 \cdot 10^{-17}$ erg.cm^{-3}.Hz^{-1} (Siscose, Davies, Coleman, Smith, Jones[186]).

For VLF and LF waves the following data are obtained

$$W_\perp \approx 10^{-17} - 10^{20} \ \text{erg/cm}^3, \qquad (3.26)$$

for longitudinal waves

$$W_\parallel \approx 10^{-13} - 10^{-16} \ \text{erg/cm}^3. \qquad (3.27)$$

In the transition zone (magnetosheath) in the frequency interval $f \approx 3-300$ Hz, which here corresponds basically to the range of VLF and LF waves, the energy density varied (see ref. 185) with increasing frequency in proportion to f^{-3} in the range $W_\perp \approx 10^{-16} - 10^{-23}$ erg.cm^{-3}. Hz^{-1}.

The HF waves have energy densities

$$W_\parallel \sim W_\perp \sim 10^{-14} - 10^{-15} \ \text{erg/cm}^3. \qquad (3.28)$$

In individual cases, however, intensive emissions have been observed with energy densities that varied in much

wider ranges than those given in (3.24)-(3.28). Such
a case is noted, for example, in §15, when the energy
density of the ion-acoustic waves was $7 \cdot 10^{-8}$ erg/cm^3
and was ~$N\kappa T$. Unfortunately, it was not always possible to determine the values of $W_{\parallel}$ and $W_{\perp}$ per hertz,
since the relevant papers do not contain data on the
effective band width Δf of the instruments employed.

A general brief conclusion that can be drawn here
is that the intensity of the longitudinal waves is usually greater, frequently by several orders of magnitude, than the intensity of the transverse waves.

BIBLIOGRAPHY

1. V. P. Silin and A. A. Rukhadze, Electromagnetic Properties of Plasmas and Similar Media [in Russian], Atomizdat, Moscow (1961).
2. T. H. Stix, The Theory of Plasma Waves, McGraw Hill, New York (1962).
3. A. I. Akhiezer, I. A. Akhiezer, R. V. Polovin, A. G. Sitenko, and K. N. Stepanov, Collective Oscillations in a Plasma, Oxford (1967).
4. V. L. Ginzburg and A. A. Rukhadze, Waves in Magnetoactive Plasmas [in Russian], Nauka (1970); Handbuch der Physik, 49/3.
5. Ya. L. Al'pert, A. V. Gurevich, and L. P. Pitaevskii, Space Physics with Artificial Satellites, Consultants Bureau, New York (1965).
6. L. D. Landau, Zh. Eksp. Teor. Fiz., 16, 574 (1946).
7. I. B. Bernstein, Phys. Rev., 109, 10 (1958).
8. P. McKeown, Rarefied Gas Dynamics, Academic Press, 2, 315 (1962).
9. G. W. Sharp, W. B. Hanson, and D. D. McKiblin, Missiles Space Company Symposium, Cospar (1963).
10. V. E. Troy, D. B. Medved, and U. Samir, Journal Astr. Sci., 18, 173 (1970).
11. Ya. L. Al'pert, Izv. Akad. Nauk SSSR Ser. Fiz., 12, 241 (1948).
12. Ya. L. Al'pert, Space. Sci. Rev., 6, 781 (1967).
13. L. R. Storey, Phil. Trans. Roy. Soc., A246, 113 (1953).
14. Ya. L. Al'pert, Usp. Fiz. Nauk, 90, 405 (1966); Space Sci. Rev., 6, 419 (1967).
15. Ya. L. Al'pert, Propagation of Electromagnetic Waves and the Ionosphere, Consultants Bureau, New York (1972).
16. Ya. L. Al'pert, A. V. Gurevich, and L. P. Pitaevskii, Usp. Fiz. Nauk, 79, 233 (1963); Space Sci. Rev., 2, 680 (1963).
17. Ya. L. Al'pert, Geomagn. Aeronom, 5, 3 (1965); Space Sci. Rev., 4, 373 (1965).
18. A. V. Gurevich, L. P. Pitaevskii, and V. V. Smirnova, Usp. Fiz. Nauk., 99, 3 (1969); Space Sci. Rev., 9, 805 (1969).
19. V. C. Liu, Space Sci. Rev., 9, 423 (1969).

20. C. L. Brundin (Ed.), Rarefied Gas Dynamics, Academic Press (1967).
21. S. F. Singer (ed.), Interaction of Space Vehicles with an Ionized Atmosphere, Pergamon Press (1965).
22. A. V. Gurevich, Trudy IZMIRAN, 17(27), 173 (1960); in: Artificial Satellites of the Earth, No. 7 [in Russian] (1961), p.101.
23. L. P. Pitaevskii Geomagn. Aeronom., $\underline{1}$, 194 (1961).
24. V. V. Vas'kov, Zh. Eksp. Teor. Fiz., $\underline{50}$, 1124 (1966).
25. N. I. Bud'ko, Zh. Eksp. Teor. Fiz., $\underline{57}$, 687 (1969).
26. A. P. Dubovoi, Zh. Eksp. Teor. Fiz., $\underline{63}$, 951 (1972).
27. A. M. Moskalenko, Candidate's Dissertation [in Russian], Moscow (1965).
28. I. A. Bogashchenko, A. V. Gurevich, and R. A. Salimov, and Yu. I. Éidel'man, Preprint, IYaF 15-70, Institute of Nuclear Physics (1970); Zh. Eksp. Teor. Fiz., $\underline{59}$, 1540 (1970).
29. W. Sawchuk, in: Rarefied Gas Dynamics, Academic Press, $\underline{2}$, 33 (1963).
30. R. L. Bowen, L. F. Boyd, C. L. Henderson, and A. P. Willmore, Proc. Roy. Soc., $\underline{A281}$, 514 (1964).
31. U. Samir and A. P. Willmore, Planet. Space Sci., $\underline{13}$, 285 (1965).
32. C. L. Henderson and U. Samir, Planet. Space Sci., $\underline{15}$, 1499 (1967).
33. W. A. Clyden and C. V. Hurdle, in: Rarefied Gas Dynamics, Academic Press (1967), p. 1717.
34. P. J. Barrett, Phys. Rev. Lett., $\underline{13}$, 742 (1964).
35. U. Samir and G. L. Wrenn, Planet. Space Sci., $\underline{17}$, 693 (1969).
36. V. V. Skvortsov and L. V. Nosachev, Kosm. Issl., $\underline{6}$, 228 (1968).
37. N. I. Bud'ko, Candidate's Dissertation [in Russian], Moscow (1969).
38. D. F. Hall, R. F. Kemp, and J. M. Sellen, AIAI Journal, $\underline{2}$, 1032 (1964).
39. A. M. Moskalenko, Geomagn. Aeronom., $\underline{4}$, 261, 509 (1964).
40. S. D. Hester and A. A. Sonin, AIAA Journal, $\underline{8}$, 1090 (1970).
41. V. C. Liu and H. Jew, in: Rarefied Gas Dynamics, Academic Press (1967), p. 1703.
42. Yu. M. Panchenko and L. P. Pitaevskii, Geomagn. Aeron. $\underline{4}$, 256 (1964).
43. Yu. M. Panchenko, Investigations of Cosmic Space [in Russian], Nauka, Moscow (1965), p. 254.
44. L. Kraus and K. Watson, Phys. Fluids, $\underline{1}$, 480 (1958).
45. L. P. Pitaevskii and V. Z. Kresin, Zh. Eksp. Teor. Fiz., $\underline{40}$, 271 (1961).

46. N. I. Bud'ko, Geomagn. Aeron., 6, 1008 (1966).
47. V. V. Vas'kov, Geomagn. Aeron., 6, 1104 (1966).
48. A. V. Gurevich, Geomagn. Aeron., 4, No. 1, 3 (1964).
49. A. V. Gurevich, Geomagn. Aeron., 3, 1021 (1963).
50. A. M. Moskalenko, Geomagn. Aeron., 4, 3 (1964).
51. V. S. Knyazyuk and A. M. Moskalenko, Geomagn. Aeron. 6, 997 (1966).
52. A. M. Moskalenko, Zh. Eksp. Teor. Fiz., 57, 1790 (1969); Geomagn. Aeron., 10, 974 (1970).
53. Ya. L. Al'pert, Usp. Fiz. Nauk, 71, 369 (1960).
54. Ya. L. Al'pert and L. P. Pitaevskii, Geomagn. Aeron. 1, 709 (1961).
55. A. V. Gurevich and L. P. Pitaevskii, Geomagn. Aeron. 9, 847 (1966).
56. V. V. Vas'kov, Geomagn. Aeron., 9, 847 (1969).
57. J. D. Kraus, R. C. Higgy, D. J. Scherr, and W. R. Crone, Nature, 185, 220 (1960).
58. J. D. Kraus, in: Interactions of Space Vehicles with an Ionized Atmosphere (Ed. S. F. Singer), Pergamon Press (1965).
59. L. D. Landau and E. M. Lifshitz, Electrodynamics of Continuous Media, Pergamon Press, Oxford (1960).
60. L. P. Pitaevskii, Geomagn. Aeron., 3, 1036 (1963).
61. V. V. Vas'kov, Kosm. Issl., 7, 559 (1969).
62. V. V. Vas'kov, Geomagn. Aeron., 7, 426 (1969).
63. V. V. Vas'kov, Candidate's Dissertation [in Russian] (1969).
64. R. A. Helliwell, VLF Observations, Report on IAGA, Moscow, 1 (1971).
65. N. I. Bud'ko, Geomagn. Aeron., 9, 430 (1969).
66. N. M. Brice and R. L. Smith, Journ. Geophys. Res., 70, 71 (1965).
67. D. J. McEwen and R. E. Barrington, Can. Journ. Phys., 45, 13 (1967).
68. A. V. Gurevich, Geomagn. Aeron., 4, 247 (1964).
69. C. T. Russell, R. E. Holzer, and E. J. Smith, Journ. Geophys. Res., 74, 755 (1969).
70. H. Benioff, Journ. Geophys. Res., 65, 1413 (1960).
71. V. A. Troitskaya, Journ. Geophys. Res., 66, 5 (1961).
72. T. Saito, Sci. Rep. Tokyo Univ., 5, (14), 81 (1962).
73. L. R. Tepley, Journ. Geophys. Res., 66, 1651 (1961).
74. L. R. Tepley and R. C. Wentworth, Journ. Geophys. Res. 67, 3312 (1962).
75. R. Gendrin and R. Stefant, Compt. Rend., 255, 752 (1962).
76. J. S. Mainstone and R. W. McNicol, Proc. Innosph. Conf., Physical Society, London (1962), p. 163.
77. J. A. Jacobs and T. Watanabe, Plan. Space. Sci., 11, 869 (1963).

78. H. W. Campbell and O. Stilner, Radio Science, Journ. Res. N.B.S., 69D, 1089 (1965).
79. J. F. Kenney and H. B. Knafich, Boeing Sci. Res. Lab., Geo-Astrophysical Laboratory Review (1968), p. 62.
80. H. B. Liemohn, Boeing Sci. Res. Lab., Document D1-82-0890 (1969).
81. J. F. Kenney, H. B. Knafich, and H. B. Liemohn, Boeing Sci. Res. Lab., Document D1-82-0691 (1968).
82. Y. Higuchi and J. A. Jacobs, Journ. Geophys. Res., 75, 7105 (1970).
83. Y. Obayshi, Jour.-Geophys. Res., 70, 1069 (1965).
84. B. Hultqvist, Space. Sci. Rev., 5, 599 (1966).
85. H. B. Liemohn, Boeing Sci. Res. Lab. Document D1-82-1043 (1971).
86. J. H. Pope, Journ. Geophys. Res., 69, 399 (1964).
87. N. A. Tartaglia, Irregular Geomagnetic Micropulsations Associated with Geomagnetic Bays in the Auroral Zone, University of Pittsburgh (1970).
88. R. L. Smith, N. M. Brice, J. Katsufrakis, D. A. Gurnett, S. D. Shawhan, T. S. Belrose, and R. E. Barrington, Nature 204, 274 (1964).
89. R. A. Helliwell, Whistlers and Related Phenomena, Stanford University Press, Stanford, Calif. (1965).
90. R. Gendrin, Hanbuch der Physik, Band 49/3, p. 461.
91. D. A. Gurnett and N. M. Brice, Journ. Geophys. Res., 71, 3639 (1966).
92. R. E. Barrington, J. S. Belrose, and W. E. Mather, Nature, 210, 80 (1966).
93. D. A. Gurnett and P. Rodriguez, Journ. Geophys. Res., 75, 1342 (1970).
94. D. A. Gurnett, S. D. Shawhan, N. M. Brice, and R. L. Smith, Journ. Geophys. Res., 70, 1665 (1965).
95. S. D. Shawhan, Journ. Geophys. Res., 71, 29 (1966).
96. S. D. Shawhan and D. A. Gurnett, Journ. Geophys. Res., 71, 47 (1966).
97. D. A. Gurnett and S. D. Shawhan, Journ. Geophys. Res., 71, 741 (1966).
98. D. A. Gurnett and N. M. Brice, Journ. Geophys. Res., 71, 3639 (1966).
99. C. Lucas and N. Brice, Journ. Geophys. Res., 76, 92 (1971).
100. W. W. Taylor and D. A. Gurnett, Journ. Geophys. Res., 73, 5615 (1968).
101. D. A. Gurnett and T. B. Burns, Journ. Geophys. Res., 73, 7437 (1968).
102. S. R. Mosier and D. A. Gurnett, Journ. Geophys. Res., 74, 5675 (1969).
103. S. R. Mosier, Journ. Geophys. Res., 76, 1713 (1971);

the University of Iowa, Document U. of Iowa 70; 2.
104. D. A. Gurnett, S. R. Mosier, and R. R. Anderson,
     Journ. Geophys. Res., 76, 3022 (1971).
105. J. L. Muzzio, Journ. Geophys. Res., 73, 7526 (1968).
106. P. Rodriguez and D. A. Gurnett, Journ. Geophys.
     Res., 76, 960 (1971).
107. D. A. Gurnett and B. J. O'Brien, Journ Geophys.
     Res., 69, 65 (1964).
108. D. A. Gurnett, C. W. Pfeiffer, R. R. Anderson, S.
     R. Mosier, and D. P. Caufman, Jour. Geophys. Res.,
     74, 4631 (1969).
109. S. R. Mosier and D. A. Gurnett, Nature, 223, N 5206,
     605 (1969).
110. D. A. Gurnett and S. R. Mosier, Journ. Geophys. Res.,
     74, 3979 (1969).
111. R. J. Stefant, Journ. Geophys. Res., 75, 7182 (1970).
112. H. Guthart, T. L. Crystal, B. P. Ficklin, W. E.
     Blair, and T. J. Yung, Journ. Geophys. Res., 73,
     3592 (1968).
113. N. Rostoker, Nuclear Fusion, 1, 101 (1961).
114. E. L. Scarf, G. M. Crook, and R. W. Fredericks,
     Journ. Geophys. Res., 70, 3045 (1965).
115. F. L. Scarf, R. W. Fredericks, and G. M. Crook,
     Journ. Geophys. Res., 73, 1723 (1968).
116. S. D. Shawhan and D. A. Gurnett, Journ. Geophys.
     Res., 73, 5649 (1968).
117. F. L. Scarf, G. M. Crook, I. M. Green, and P. F.
     Virobik, Journ. Geophys. Res., 73, 6665 (1968).
118. R. E. Barrington and J. S. Belrose, Nature, 198,
     651 (1963).
119. R. E. Barrington, J. S. Belrose, and D. A. Keeley,
     Journ. Geophys. Res., 68, 6539 (1963).
120. R. E. Barrington, J. S. Belrose, and C. L. Nelms,
     Journ. Geophys. Res., 70, 1647 (1965).
121. N. M. Brice and R. L. Smith, Journ. Geophys. Res.,
     70, 71 (1965).
122. T. Laaspere, M. G. Morgan, and W. C. Johnson, Journ.
     Geophys. Res., 74, 141 (1969).
123. T. Laaspere and H. A. Taylor, Journ Geophys. Res.,
     75, 97 (1970).
124. E. L. Scarf, R. W. Fredericks, E. J. Smith, A. M
     Frandsen, and G. P. Serbu, Space Science Department,
     California, Document 05402-6031-R0-OO, August (1971).
125. D. A. Gurnett and L. A. Frank, University of Iowa,
     Document U. of Iowa 71:19.
126. S. R. Mosier and D. A. Gurnett, University of Iowa,
     Document U. of Iowa 71:21.
127. R. E. Barrington and J. S. Belrose, Nature, 198,
     651 (1963).

128. D. L. Carpenter, N. Dunckel, and J. F. Walcup, Journ. Geophys. Res., 69, 5009 (1964).

129. C. O. Hines, Journal. Atm. Terr. Phys., 11, 36 (1957).

130. R. L. Smith, Journ. Geophys. Res., 69, 5019 (1964).

131. I. Kimura, Radio Science (New Series), 1, 269 (1966).

132. R. M. Thorne and C. F. Kennel, Journ. Geophys. Res., 72, 857 (1967).

133. D. L. Carpenter and N. Dunckel, Journ. Geophys. Res., 70, 3781 (1965).

134. I. Kimura, R. L. Smith. and N. M. Brice, Journ Geophys. Res., 70, 5961 (1965).

135. R. L. Smith and J. J. Angerami, Journ. Geophys. Res., 73, 1 (1968).

136. W. C. Hoffman, Journ. Atm. and Terr. Phys. 18, 1 (1960).

137. R. M. Thorne, Journ. Geophys. Res., 73, 4895 (1968).

138. D. L. Carpenter, Journ. Geophys. Res., 68, 1675 (1963).

139. D. L. Carpenter, Journ. Geophys. Res., 71, 693 (1966).

140. J. J. Angerami and D. L. Carpenter, Journ. Geophys. Res., 71, 711 (1966).

141. D. L. Carpenter, C. G. Park, H. A. Taylor, and N. C. Brinton, Journ. Geophys. Res., 74, 1837 (1969).

142. D. L. Carpenter, Journ. Geophys. Res., 73, 2919 (1968).

143. R. L. Smith, R. A. Helliwell, and I. W. Yabroff, Journ. Geophys. Res., 65, 815 (1960).

144. R. L. Smith, Journ. Geophys. Res., 66, 3699 (1961).

145. J. J. Angerami, Journ. Geophys. Res., 75, 6115 (1970).

146. W. J. Burtis and R. A. Helliwell, Journ. Geophys. Res., 74, 3002 (1969).

147. N. Dunckel and R. A. Helliwell, Journ. Geophys. Res., 74, 6371 (1969).

148. N. Dunckel, B. Ficklin, R. Rorden, and R. A. Helliwell, 75, 1854 (1970).

149. C. F. Kennel and H. E. Petschek, Journ. Geophys. Res., 71, 1 (1966).

150. F. V. Coroniti, R. E. Fredericks, C. F. Kennel, and F. L. Scarf, Journ. Geophys. Res., 76, 2366 (1971).

151. Proceedings of the IEEE, 57, N6 (1969).

152. W. Calvert and J. R. McAfee, Proc. IEEE, 57, 1089 (1969).

153. C. E. Lockwood, Can. Journ. Phys. 43, 291 (1965).

154. G. L. Nelms and G. E. Lockwood, Space Res., 7, 604 (1966).

155. H. Oya, Journ. Geophys. Res., 75, 4279 (1970).

156. E. S. Warren and E. L. Hagg, Nature, 22, 1968 (1968).

157. J. P. Dougherty and J. J. Monaghan, Proc. Roy. Soc.,
     A.289, 214 (1965).
158. J. A. Fejer and W. Calvert, Journ. Geophys. Res.,
     69, 5049 (1964).
159. F. W. Crawford, R. S. Harp, and T. D. Mantei, Journ.
     Geophys. Res., 72, 57 (1967).
160. C. F. Kennel, F. L. Scarf. R. W. Fredericks, J. H.
     McGehee, and F. V. Coroniti, Journ. Geophys. Res.,
     75, 6136 (1970).
161. F. L. Scarf and R. W. Fredericks, Space Science
     Department California, Document 05402-6030-R0-00,
     August (1971).
162. F. L. Scarf, R. W. Fredericks, I. M. Green, and M.
     Neugebauer, Journ. Geophys. Res., 75, 3735 (1970).
163. G. L. Siscose, F. L. Scarf, I. M. Green, J. H. Bin-
     sack, and H. S. Bridge, Journ. Geophys. Res., 76,
     828 (1971).
164. F. L. Scarf, R. W. Fredericks, and I. M. Green,
     Journ. Geophys. Res., Space Science Laboratories,
     California, Report No. 17706-6002-R0-00 March (1971).
165. U. Samir, Israel Journal of Technology, 10, 179
     (1972).
166. U. Samir and G. L. Wrenn, Plan. Space Sci., 20, 899
     (1972).
167. E. J. Smith, R. E. Holzer, and C. T. Russell, Journ.
     Geophys. Res., 74, 3027 (1969).
168. D. A. Gurnett and L. A. Frank, Journ. Geophys. Res.,
     77, 3411 (1972).
169. A. V. Gurevich, R. A. Salimov, and N. S. Buchel'-
     nikova, Teplofiz. Vys. Temp., 7, 852 (1969).
170. V. T. Astrelin, I. A. Bogashchenko, N. S. Buchel'-
     nikova, and Yu. I. Éidel'man, Preprints IYaF 41-71
     (1971), IYaF 9-72 (1972), IYaF 13-72 (1972), Insti-
     tute of Nuclear Physics; Zh. Tekh. Fiz., 42, 1715
     (1972).
171. J. T. M. Schmitt, Laboratoire des Physique des
     Milieaux Ionise, Ecole Politeqnique, Paris, France
     (1972).
172. Yu. S. Sayasov and L. A. Zhizhimov, Radiotekh. El-
     ektron., 8, 499 (1963).
173. V. V. Smirnova, Geomagn. Aeron., 7, 33 (1967).
174. V. V. Smirnova, Zh. Tekh. Fiz., 39, 49 (1969).
175. A. V. Gurevich, L. V. Pariiskaya, and L. P. Pitaev-
     skii, Zh. Eksp. Teor. Fiz., 63, 516 (1972).
176. R. R. Shaw and D. A. Gurnett, The University of
     Iowa, U. of Iowa 72:37, December (1972).
177. R. A. Helliwell, Journ. Geophys. Res., 68, 5387
     (1963).
178. R. A. Helliwell, J. Katsufrakis, M. Trimpi, and

N. Brice, Journ. Geophys. Res., $\underline{69}$, 2391 (1964).

179. R. A. Helliwell, J. Katsufrakis, and J. Kimura,
     Paper presented at IRE URSI Symposium April (1965).

180. I. Kimura, Journ. Geophys. Res., $\underline{73}$, 445 (1968).

181. R. A. Helliwell, Journ. Geophys. Res., $\underline{72}$, 4773
     (1967).

182. I. Kimura, Planet. Space Sci. $\underline{15}$, 1462 (1967).

183. H. Matsumoto, Theoretical Studies on Whistler Mode
     Wave-Particle Interactions in the Magnetospheric
     Plasma, December (1972), Kyoto University, Japan.

184. P. J. Coleman, Journ. Geophys. Res., $\underline{69}$, 3051 (1964).

185. E. J. Smith, R. E. Holzer, M. G. McLeod, and C. T.
     Russell, Journ. Geophys. Res., $\underline{72}$, 4803 (1967).

186. G. L. Siscose, L. Davies, P. J. Coleman, E. J. Smith,
     and D. E. Jones, Journ. Geophys. Res., $\underline{72}$, 1 (1967);
     $\underline{73}$, 61 (1968).